I0475918

The Quantum Frontier:

Exploring the Weird and Wonderful World of
Quantum Physics

Written, Edited, and Illustrated by

Nicholas Tassy-Becz

Dedicated to my amazing nana

Colleen Jewett

(about as weird and wonderful as quantum physics)

ISBN: 978-1-329-19896-8

Nicholas Tassy-Becz

Preface

In this book, we will mostly be talking about quantum physics. Still, I hope that I will convey the information to you in a more digestible way than an average textbook with dictation and explanation that is accessible to persons of all backgrounds. That being said we will cover math that is unfortunately complex. Therefore the targeted reader of this book is someone with a general knowledge of topics like linear algebra and complex geometry. If you have not yet familiarized yourself with these topics, please disregard the math portions and stick to the intriguing science sections. Over the course of this book, we will delve into deeper and deeper aspects of quantum physics and by extension get into more present, frontiers of science. If you become lost in the complexities of this admittedly very challenging subject feel no shame in skipping chapters. For reference, the most comprehensible material will be found in chapters 1, 9, and 10. In order of complexity comes 2-8. That being said, I hope you enjoy your reading and if you do complete this book with a general understanding of quantum physics consider yourself a smart cookie.

Nicholas Tassy-Becz

Table of Context

Chapter 1: The Foundations

In this chapter, we will explore the foundational concepts of quantum physics, including the wave-particle duality, the uncertainty principle, and the superposition principle. We will learn about the experiments that led to the development of these concepts and how they challenge our classical understanding of physics.

Quantum physics is the branch of physics that studies the behavior of matter and energy at the smallest scales, such as atoms, subatomic particles, and photons. It is a fundamental theory that describes the strange and counterintuitive behavior of these tiny particles, such as their ability to exist in multiple states simultaneously, to be entangled with each other across vast distances, and to exhibit unpredictable behavior that defies our classical understanding of the world. Quantum physics has revolutionized our understanding of the universe and has led to numerous technological advances, including the development of lasers, transistors, and quantum computers.

One clarification to make is the subject of observation, when the term 'observed' is used we like to think of it as an abstract philosophical concept but in reality, all this means is that a photon touched the particle. A way to think of this is to see something we need to use light (as our eyes can't see without it) but that light when touching extremely small particles will change them so the ways we literally see them and the way we can observe their effects are different.

As stated previously there are three fundamental concepts that one needs to understand before we progress through harder topics and those are wave-particle duality, the uncertainty principle, and the superposition principle.

Wave-particle duality is a fundamental concept in quantum mechanics that states that particles, such as electrons and photons, exhibit both wave-like and particle-like behavior. This concept challenges the classical understanding of physics, where particles were considered to be discrete, point-like entities that followed predictable paths.

The wave-particle duality was first observed in the famous double-slit experiment, conducted by Thomas Young in the early 19th century. In this experiment, a beam of light was passed through two slits and projected onto a screen. The interference pattern that was observed on the screen suggested that the light behaved like a wave.

This wave-like behavior of light was further supported by the experiments conducted by James Clerk Maxwell in the mid-19th century. Maxwell's equations predicted the existence of electromagnetic waves, which were later confirmed by experiments conducted by Heinrich Hertz.

The wave-particle duality was further confirmed by the experiments conducted by Louis de Broglie in the early 20th century. de Broglie proposed that particles, such as electrons, could also exhibit wave-like behavior, with their wavelength being inversely proportional to their momentum. This was confirmed by the diffraction experiments conducted by Davisson and Germer in 1927, which showed that electrons could exhibit interference patterns similar to those observed in the double-slit experiment.

The wave-particle duality is also central to the understanding of the uncertainty principle, which states that the position and momentum of a particle cannot be known simultaneously with arbitrary precision. This is because the act of measurement necessarily affects the state of the particle, making it impossible to know both values with perfect accuracy.In the case of a free particle, the wave function takes the form of a plane wave, which is described by the de Broglie relation:

$$p = \hbar k$$

where p is the momentum of the particle, k is the wave vector, and ħ is the reduced Planck constant. The wave function of a free particle is given by:

$$\Psi(x,t) = A\exp[i(kx - \omega t)]$$

where A is the amplitude of the wave, k is the wave vector, ω is the frequency, and x and t are the position and time of the particle, respectively.

The wave function also obeys the Schrödinger equation, which describes the time evolution of the wave function. The Schrödinger equation is given by:

$i\hbar \partial \Psi / \partial t = H\Psi$

where $\hbar$ is the reduced Planck constant, Ψ is the wave function, and H is the Hamiltonian operator, which describes the energy of the system. The Schrödinger equation is a partial differential equation that describes the time evolution of the wave function.

The wave-particle duality principle also implies that particles can be described by wave packets, which are localized wave functions that describe the probability density of the particle's position. The wave packet is obtained by superimposing plane of waves with different wave vectors and frequencies. The wave packet is given by:

$$\Psi(x,t) = \int \Phi(k) \exp[i(kx - \omega t)] dk$$

where $\Phi(k)$ is the Fourier transform of the wave function, and dk is the differential of the wave vector. The wave packet describes the probability density of the particle's position and Momentum, and it exhibits both wave-like and particle-like behavior.

The wave-particle duality principle also implies that particles can be described by a probability distribution, which is given by the squared modulus of the wave function. The probability density of finding the particle at a specific position x at a specific time t is given by:

$$|\Psi(x,t)|^2 = \Psi^*(x,t)\Psi(x,t)$$

where Ψ^* is the complex conjugate of the wave function. The probability density is a real-valued function that describes the probability of finding the particle at a specific position.

In conclusion, the mathematical formalism behind wave-particle duality is essential to understanding and describing the phenomenon. The wave function describes the state of a quantum system, and the Schrödinger equation describes the time evolution of the wave function. The wave-particle duality principle implies that particles can exhibit both wave-like and particle-like behavior, and the wave function, wave packets, and probability density are used to describe this behavior mathematically. The mathematical formalism behind wave-particle duality is a fundamental component of quantum mechanics and has significant implications for our understanding of the quantum world.

The Uncertainty Principle in Quantum Physics

The uncertainty principle is a fundamental concept in quantum mechanics that was first introduced by Werner Heisenberg in 1927. It states that the position and momentum of a particle cannot be known simultaneously with arbitrary precision. This principle challenges the classical understanding of physics, where the position and momentum of a particle were believed to be well-defined and predictable.

The uncertainty principle can be expressed mathematically as follows:

$$\Delta x \Delta p \geqslant \hbar/2$$

where Δx represents the uncertainty in the position of the particle, Δp represents the uncertainty in the momentum of the particle, and $\hbar$ is the reduced Planck constant.

This mathematical expression implies that the product of the uncertainties in position and momentum must be greater than or equal to a constant value, which is on the order of Planck's constant. Therefore, the more precisely we measure the position of a particle, the less precisely we can measure its momentum and vice versa

Moreover, the uncertainty principle suggests that the behavior of particles at the quantum level is probabilistic, rather than deterministic. This means that the exact trajectory of a particle cannot be predicted with absolute certainty and that the behavior of a particle can only be described in terms of probabilities.

The uncertainty principle is not limited to the position and momentum of a particle. It also applies to other complementary pairs of physical quantities, such as energy and time. The mathematical expression for this complementary pair of variables is as follows:

$$\Delta E \Delta t \geqslant \hbar/2$$

where ΔE represents the uncertainty in the energy of a system, Δt represents the uncertainty in the time interval over which the measurement is made, and $\hbar$ is the reduced Planck constant.

The uncertainty principle has important implications for a wide range of phenomena in quantum mechanics, such as the behavior of atoms, the structure of the nucleus, and the behavior of subatomic particles. It is also essential to the understanding of quantum computing, where the principle plays a central role in the design and operation of quantum algorithms.

In conclusion, the uncertainty principle is a fundamental concept in quantum mechanics that challenges the classical understanding of physics. It implies that the position and momentum of a particle cannot be known simultaneously with arbitrary precision, and that the behavior of particles at the quantum level is probabilistic. The principle has important implications for a wide range of phenomena in quantum mechanics and plays a central role in the design and operation of quantum computing.

The Superposition Principle

The superposition principle is a fundamental concept in quantum mechanics that states that a quantum system can exist in multiple states simultaneously. This principle challenges the classical understanding of physics, where a system was believed to exist in a single, well-defined state.

The superposition principle can be expressed mathematically as follows:

$$|\psi\rangle = c1|\psi1\rangle + c2|\psi2\rangle + \dots$$

where $|\psi\rangle$ represents the state of the system, c1, c2, ... are complex numbers known as coefficients, and $|\psi1\rangle$, $|\psi2\rangle$, ... represent the basis states of the system.

This mathematical expression implies that the state of a quantum system can be expressed as a linear combination of basis states, each with a complex coefficient. The coefficients determine the probability amplitudes for each basis state, which are used to calculate the probability of observing a particular outcome when a measurement is made.

The superposition principle is also essential to the understanding of quantum computing, where the principle is used to represent and manipulate quantum information. Quantum computers use quantum bits, or qubits, which can exist in a superposition of states, to perform calculations in parallel and solve problems that are difficult or impossible to solve using classical computers.

In conclusion, the superposition principle is a fundamental concept in quantum mechanics that challenges the classical understanding of physics. It suggests that a quantum system can exist in multiple states simultaneously, and that the behavior of particles at the quantum level is probabilistic. The principle has important implications for the behavior of particles at the quantum level and plays a central role in the design and operation of quantum computing.

Particle accelerators are powerful scientific tools that have played a critical role in advancing our understanding of the universe. These machines use electric and magnetic fields to accelerate charged particles, such as protons or electrons, to incredibly high speeds, often approaching the speed of light. This acceleration allows scientists to study the properties of these particles and the interactions between them.

There are two main types of particle accelerators: linear accelerators (linacs) and circular accelerators. Linacs use a series of electric fields to accelerate particles in a straight line, while circular accelerators use magnetic fields to bend the path of the particles into a circular orbit.

The earliest particle accelerators were built in the early 20th century and were used primarily for studying the properties of atoms. Today, accelerators have a wide range of scientific and practical applications, including fundamental research in particle physics, medical imaging and therapy, and materials science.

One of the most well-known accelerators is the Large Hadron Collider (LHC) located at CERN in Switzerland. The LHC is a circular accelerator that uses powerful superconducting magnets to steer and focus two beams of protons in opposite directions. When the two beams collide, the energy released can create new particles that have never been observed before, allowing physicists to study the fundamental properties of matter.

Another important application of particle accelerators is in cancer treatment. In this case, the accelerator is used to generate a beam of high-energy particles, which are directed at the tumor to destroy cancerous cells while minimizing damage to healthy tissue. This technique, known as radiation therapy, has become an important tool in cancer treatment and has saved countless lives.

In addition to their practical applications, particle accelerators also have the potential to advance our understanding of the universe in profound ways. For example, accelerators can be used to recreate the conditions that existed in the early universe just after the Big Bang, allowing scientists to study the fundamental forces of nature and the origins of matter.

Despite their many benefits, particle accelerators also pose some challenges. For example, building and operating these machines can be incredibly expensive, with some accelerators costing billions of dollars to construct and maintain. Additionally, the high energies involved in accelerator experiments can be dangerous, requiring stringent safety measures and protocols to ensure the well-being of both researchers and the public.

In conclusion, particle accelerators are powerful tools that have revolutionized our understanding of the universe and have numerous practical applications. From fundamental research in particle physics to cancer treatment, these machines have the potential to transform our world in countless ways. As scientists continue to push the boundaries of what is possible with these machines, it is likely that we will see even more exciting discoveries and applications emerge in the years to come.

Alpha, beta, and gamma decay are three types of radioactive decay that occur in unstable nuclei. These processes result in the emission of different types of radiation and are important for understanding the properties and behavior of radioactive elements.

Alpha decay occurs when an atomic nucleus emits an alpha particle, which consists of two protons and two neutrons bound together. This emission reduces the atomic number of the nucleus by two and the mass number by four. Alpha decay is most commonly observed in heavy elements such as uranium, thorium, and radium. This type of decay is relatively slow and can be stopped by a sheet of paper or a few centimeters of air.

Beta decay occurs when a nucleus emits a beta particle, which is either a high-energy electron or a positron (an antimatter particle with the same mass as an electron but with a positive charge). Beta decay can result in the conversion of a neutron to a proton or a proton to a neutron, which changes the atomic number of the nucleus but not its mass number. Beta decay is faster than alpha decay and can be stopped by a few millimeters of aluminum or other dense material.

Gamma decay is the emission of a gamma ray, which is a high-energy photon. Gamma rays are emitted when a nucleus transitions from a higher energy state to a lower energy state. Unlike alpha and beta particles, gamma rays do not carry a charge or have a mass, so they can penetrate further through matter. Gamma decay does not change the atomic or mass number of the nucleus and can only be stopped by several centimeters of dense material, such as lead.

Each type of decay has its own characteristics, and the properties of the emitted radiation can provide important information about the properties of the nucleus undergoing decay. For example, the emission of an alpha particle is indicative of a heavy, unstable nucleus, while the emission of a gamma ray can provide information about the energy levels of the nucleus.

Radioactive decay has many important applications, including in nuclear power and medicine. In nuclear power, the heat generated by the decay of radioactive elements is used to produce electricity. In medicine, radioactive isotopes can be used for diagnostic imaging and cancer treatment.

In conclusion, alpha, beta, and gamma decay are three types of radioactive decay that occur in unstable nuclei. Each type of decay results in the emission of different types of radiation and has its own unique characteristics. Understanding the properties of radioactive decay is essential for understanding the behavior of radioactive elements and for developing applications in fields such as nuclear power and medicine.

Chekov radiation, also known as Vavilov-Cherenkov radiation, is a type of electromagnetic radiation that is emitted when a charged particle travels through a medium at a speed greater than the phase velocity of light in that medium. This phenomenon is named after the Russian physicist Pavel Alekseyevich Cherenkov, who first observed it in 1934, and the Russian physicist Sergey Ivanovich Vavilov, who predicted its existence in 1934.

When a charged particle, such as an electron or a proton, moves through a medium, it causes the atoms in the medium to become polarized, creating a disturbance in the electric field. If the particle is moving faster than the phase velocity of light in the medium, this disturbance causes the emission of radiation in a cone-shaped pattern that is centered on the direction of motion of the particle. The angle of the cone is given by the ratio of the speed of the particle to the speed of light in the medium, and the intensity of the radiation is proportional to the square of the charge of the particle and the square of its velocity.

Chekov radiation is important in a wide range of fields, including particle physics, astrophysics, and medical imaging. In particle physics, the radiation is used to identify and measure the properties of high-energy charged particles, such as those produced in particle accelerators. In astrophysics, the radiation is used to study the properties of high-energy cosmic rays, as well as the properties of high-energy particles emitted by stars and other celestial objects. In medical imaging, the radiation is used in a technique known as positron emission tomography (PET), which allows for the detection and imaging of the distribution of radioactive isotopes within the body.

One of the most important applications of Chekov radiation is in the detection of neutrinos, which are high-energy, electrically neutral particles that are produced in nuclear reactions and other astrophysical phenomena. Neutrinos interact very weakly with matter and are difficult to detect directly, but when a neutrino interacts with an atomic nucleus, it can produce a charged particle that travels faster than the speed of light in the medium. This produces a burst of Chekov radiation that can be detected by sensitive instruments, allowing for the detection of neutrinos and the study of their properties.

In conclusion, Chekov radiation is a type of electromagnetic radiation that is emitted when a charged particle travels through a medium at a speed greater than the phase velocity of light in that medium. This phenomenon has important applications in fields such as particle physics, astrophysics, and medical imaging, and it plays a crucial role in the detection and study of high-energy particles and neutrinos. The discovery of Chekov radiation has had a profound impact on our understanding of the behavior of charged particles and the properties of matter and radiation

The Planck constant is a fundamental physical constant that plays a central role in the field of quantum mechanics. It is denoted by the symbol h and has a value of approximately 6.626 x 10^-34 joule-seconds.

At its core, the Planck constant relates the energy of a photon to its frequency. In other words, it provides a way to connect the particle-like nature of light (as photons) to its wave-like properties (as a frequency). This connection is crucial to understanding the behavior of matter and energy at the atomic and subatomic level.

One way to understand the Planck constant is to consider the concept of quantization, which is central to quantum mechanics. Quantization refers to the idea that certain physical properties, such as energy, can only take on specific, discrete values. This stands in contrast to classical physics, where physical properties can take on any value within a continuous range.

The Planck constant emerges in the process of quantization when considering the energy of a photon. In quantum mechanics, the energy of a photon is given by the equation $E = hf$, where E is the energy of the photon, h is the Planck constant, and f is the frequency of the photon.

This equation shows that the energy of a photon is directly proportional to its frequency. This means that photons with higher frequencies (i.e., shorter wavelengths) have higher energy than photons with lower frequencies (i.e., longer wavelengths). This relationship is a fundamental principle of quantum mechanics and is known as the Planck-Einstein relation.

The Planck constant also appears in other equations in quantum mechanics, such as the Heisenberg uncertainty principle. This principle states that there is a limit to how accurately certain physical properties, such as position and momentum, can be measured at the same time. The limit is proportional to the Planck constant, which reflects the fundamental indeterminacy of the quantum world.

In summary, the Planck constant is a fundamental physical constant that relates the energy of a photon to its frequency. It plays a central role in the field of quantum mechanics and is essential for understanding the behavior of matter and energy at the atomic and subatomic levels. While its implications are complex and far-reaching, its basic meaning is simple: it describes the fundamental relationship between light as both particles and waves.

Unfortunately, most of what we understand about quantum physics breaks what we would consider 'normal' and that's what makes it so hard. To that note, try not to overthink everything stated and just trust that these laws are rules and are true and apply, and disregard why in the world they should.

Chapter 2: History

Trial and error is practically the only method we have in quantum physics to create rules that govern particles. In this chapter, we will cover the experiments and the incredible scientists behind them that lead to a greater understanding and perhaps discoveries that better our comprehension of such an incomprehensible field.

Quantum physics was not founded by a single individual but rather emerged from the work of many scientists in the early 20th century. However, some of the key contributors to the development of quantum physics include Max Planck, who first proposed the idea of the quantization of energy, Albert Einstein, who explained the photoelectric effect using the concept of photons, Niels Bohr, who proposed the first successful model of the atom, and Werner Heisenberg, who formulated the uncertainty principle, which describes the limitations of our ability to measure certain properties of particles. Other notable physicists who contributed to the development of quantum physics include Erwin Schrödinger, Paul Dirac, Richard Feynman, and many others.

Planck's most significant contribution to quantum physics was his proposal of the concept of quantization of energy. Prior to Planck's work, classical physics assumed that energy was continuous and could take on any value. However, Planck's analysis of blackbody radiation led him to propose that energy was

23

quantized, meaning it could only exist in discrete packets or quanta. This idea was revolutionary and laid the foundation for the development of quantum mechanics.

Max Planck was a German physicist who is widely considered one of the founders of quantum physics. His contributions to the field were groundbreaking and helped pave the way for many of the developments in modern physics. Planck's most significant contribution to quantum physics was his proposal of the concept of quantization of energy. Prior to Planck's work, classical physics assumed that energy was continuous and could take on any value. However, Planck's analysis of blackbody radiation led him to propose that energy was quantized, meaning it could only exist in discrete packets or quanta. This idea was revolutionary and laid the foundation for the development of quantum mechanics. Planck's proposal of quantization of energy was motivated by his attempts to explain the spectral distribution of energy emitted by a blackbody, which is an object that absorbs all radiation that falls on it. According to classical physics, the radiation emitted by a blackbody should have an infinite range of frequencies and energies, but experimental observations did not match this prediction. Planck proposed that energy was quantized and could only be emitted or absorbed in discrete units, which explained the observed spectral distribution of energy.

Planck's work on the quantization of energy led to the development of Planck's constant, which is a fundamental constant of nature that relates the energy of a photon to its frequency. This constant is used extensively in quantum mechanics and is critical for understanding the behavior of particles at the atomic and subatomic levels. Planck's work on the quantization of energy also had significant implications for the development of atomic theory. In particular, it inspired the development of the Bohr model of the atom, which proposed that electrons orbit the nucleus in discrete energy levels, rather than in a continuous manner as predicted by classical physics. In addition to his contributions to quantum physics, Planck also made significant contributions to other areas of physics, including thermodynamics and radiation theory. He was awarded the Nobel Prize in Physics in 1918 for his work on blackbody radiation.

In conclusion, Max Planck's contributions to quantum physics were transformative and helped establish the field as a fundamental pillar of modern physics. His proposal of the concept of quantization of energy laid the foundation for the development of quantum mechanics and had significant implications for atomic theory. Planck's work on the quantization of energy and the development of Planck's constant remains fundamental to our understanding of the behavior of particles at the atomic and subatomic level

Albert Einstein, one of the most influential scientists of the 20th century, made numerous contributions to physics, including the development of the theory of relativity and the famous equation $E=mc^2$. However, he also made significant contributions to the field of quantum physics. Einstein's work in quantum physics began with his explanation of the photoelectric effect in 1905. The photoelectric effect is the phenomenon in which electrons are emitted from a metal surface when it is exposed to light. Classical physics predicted that the energy of the emitted electrons would increase with the intensity of the light, but the observed results were inconsistent with this prediction. Einstein proposed that the energy of the emitted electrons was proportional to the frequency of the light, rather than its intensity. This explanation was a major step towards understanding the wave-particle duality of light and laid the foundation for the development of quantum mechanics. Einstein's contributions to quantum physics also include his work on the concept of wave-particle duality. In 1909, he proposed that light had both wave-like and particle-like properties, and in 1915, he extended this concept to matter by proposing that particles, such as electrons, also exhibited wave-like properties. This idea was later developed into the theory of wave mechanics, which is a fundamental principle of quantum mechanics. Another significant contribution of Einstein to quantum physics was his criticism of the Copenhagen interpretation of quantum mechanics. The Copenhagen interpretation, proposed by Niels Bohr and Werner Heisenberg, states that the act of observation of a particle affects its properties and that particles do not have definite properties until they are observed. Einstein was critical of this interpretation and proposed alternative theories, such as the hidden variable theory, which posits that particles have definite properties that are not observable. Einstein's work on the theory of relativity also had significant implications for quantum mechanics. His theory predicted that the speed of light was constant in all reference frames, which contradicted classical physics and led to the development of a new understanding of space and time. This new understanding of space and time was incorporated into the development of quantum mechanics, which has become a fundamental theory for understanding the behavior of particles at the atomic and subatomic levels.

In conclusion, Albert Einstein made significant contributions to quantum physics, including his explanation of the photoelectric effect, his work on wave-particle duality, his criticism of the Copenhagen interpretation, and the implications of his theory of relativity. His work laid the foundation for the development of quantum mechanics, which has become a cornerstone of modern physics. Einstein's contributions to physics have had a profound impact on our understanding of the universe and continue to inspire scientists today.

Niels Bohr was a Danish physicist who made several significant contributions to the field of quantum physics. His work helped lay the foundations of quantum mechanics and led to a better understanding of the behavior of particles at the atomic and subatomic levels. Bohr's most significant contribution to quantum physics was the development of the Bohr model of the atom. The Bohr model proposed that electrons orbit the nucleus of an atom in discrete energy levels, rather than in a continuous manner as predicted by classical physics. This model helped to explain the observed spectral lines of elements and was a significant step toward understanding the behavior of atoms. Bohr's model was based on the idea that electrons could only occupy certain energy levels, and that the energy of an electron was determined by the frequency of its orbit. This idea was inspired by the work of Max Planck, who proposed the concept of quantization of energy. Bohr's model of the atom had significant implications for the understanding of the atomic structure and chemical bonding. It helped to explain the periodicity of the elements and the stability of chemical compounds. The Bohr model also provided a framework for understanding the behavior of atoms in external fields, which led to the development of important technologies such as nuclear magnetic resonance imaging (MRI). In addition to his work on the Bohr model, Bohr made several other contributions to quantum physics. He proposed the principle of complementarity, which states that particles can exhibit both wave-like and particle-like behavior, but. That these behaviors cannot be observed simultaneously. This principle helped to resolve some of the inconsistencies in quantum mechanics and has become an important concept in modern physics. Bohr also developed the concept of the Copenhagen interpretation of quantum mechanics, which states that particles do not have definite properties until they are observed. This interpretation has been the subject of much debate and criticism but has remained a fundamental concept in quantum mechanics.

In conclusion, Niels Bohr made several significant contributions to quantum physics, including the development of the Bohr model of the atom, the principle of complementarity, and the Copenhagen interpretation of quantum mechanics. His work helped to establish the foundations of quantum mechanics and has had a significant impact on our understanding of the behavior of particles at the atomic and subatomic level. Bohr's legacy continues to inspire scientists and researchers today, and his contributions to physics remain a cornerstone of modern physics.

Werner Heisenberg was a German physicist who made significant contributions to the development of quantum mechanics, which is a fundamental theory for understanding the behavior of particles at the atomic and subatomic level. Heisenberg's work helped to establish the foundations of quantum mechanics and has had a significant impact on modern physics.

One of Heisenberg's most significant contributions to quantum physics was the development of the uncertainty principle. The uncertainty principle states that the more precisely one knows the position of a particle, the less precisely one can know its momentum, and vice versa. This principle challenged the classical notion of determinism and led to a new understanding of the behavior of particles at the atomic and subatomic level. Heisenberg also played a central role in the development of matrix mechanics, which is one of the three main formulations of quantum mechanics. Matrix mechanics describes the behavior of particles in terms of matrices and was developed independently by Heisenberg, Max Born, and Pascual Jordan. This formulation helped to establish the mathematical framework for quantum mechanics and has become an essential tool for studying the behavior of particles at the atomic and subatomic level. Another significant contribution of Heisenberg to quantum physics was his development of the concept of quantum field theory. Quantum field theory describes the behavior of particles as excitations of a quantum field and has become a fundamental theory for understanding the behavior of elementary particles. This theory has led to significant advances in particle physics and has helped to explain the behavior of particles in high-energy collisions. Heisenberg's work also had significant implications for the understanding of nuclear physics. He was one of the leading scientists involved in the development of nuclear weapons during World War II, and his work helped to establish the principles for the design of nuclear reactors and the production of nuclear energy.

In conclusion, Werner Heisenberg made several significant contributions to quantum physics, including the development of the uncertainty principle, matrix mechanics, and quantum field theory. His work helped to establish the foundations of quantum mechanics and has had a significant impact on our understanding of the behavior of particles at the atomic and subatomic level. Heisenberg's legacy continues to inspire scientists and researchers today, and his contributions to physics remain a cornerstone of modern physics.

Erwin Schrödinger was an Austrian physicist who made several significant contributions to the field of quantum mechanics. His work helped to establish the mathematical framework for quantum mechanics and has had a significant impact on modern physics. Schrödinger's most significant contribution to quantum physics was the development of the Schrödinger equation. The Schrödinger equation is a fundamental equation that describes the behavior of particles in terms of a wave function. It helped to establish the mathematical framework for quantum mechanics and has become an essential tool for studying the behavior of particles at the atomic and subatomic level.

Schrödinger also made significant contributions to the understanding of quantum entanglement, which is a phenomenon where the properties of two particles become correlated in a way that cannot be explained by classical physics. Schrödinger proposed the concept of entanglement in his famous thought experiment involving a cat in a box, which is known as the Schrödinger's cat paradox. This concept helped to establish the foundation for the study of quantum information and has become an important concept in modern physics. Another significant contribution of Schrödinger to quantum physics was his work on the interpretation of quantum mechanics. He proposed the concept of wave-particle duality, which states that particles can exhibit both wave-like and particle-like behavior depending on how they are observed.

Schrödinger also proposed the concept of the wave function collapse, which describes how the properties of particles become fixed when they are observed. These concepts helped to establish a more complete understanding of the behavior of particles in the quantum world. Schrödinger's work also had significant implications for the understanding of molecular structure and chemical bonding. He proposed the concept of resonance structures, which describe how electrons can be distributed across different atoms in a molecule. This concept helped to explain the observed properties of many organic molecules and has become an essential tool for understanding the behavior of molecules.

In conclusion, Erwin Schrödinger made several significant contributions to quantum physics, including the development of the Schrödinger equation, the concept of entanglement, and the interpretation of quantum mechanics. His work helped to establish the mathematical framework for quantum mechanics and has had a significant impact on our understanding of the behavior of particles at the atomic and subatomic level. Schrödinger's legacy continues to inspire scientists and researchers today, and his contributions to physics remain a cornerstone of modern physics.

Paul Dirac was an English physicist who made several significant contributions to the field of quantum mechanics. His work helped to establish the foundations of quantum mechanics and has had a significant impact on modern physics. Dirac's most significant contribution to quantum physics was the development of the Dirac equation, which describes the behavior of relativistic particles, such as electrons, in terms of a wave function. The Dirac equation helped to establish the mathematical framework for quantum field theory and has become an essential tool for studying the behavior of particles in high-energy collisions. Dirac also made significant contributions to the understanding of antimatter. In 1928, Dirac proposed the concept of the positron, which is the antiparticle of the electron. This concept helped to establish the foundations for the study of antimatter and has had significant implications for particle physics and cosmology. Another significant contribution of Dirac to quantum physics was his work on the interpretation of quantum mechanics. Dirac proposed the concept of quantum electrodynamics, which describes the behavior of particles in terms of virtual photons. This concept helped to establish the foundations for the study of particle interactions and has become an essential tool for understanding the behavior of particles in the quantum world. Dirac's work also had significant implications for the understanding of magnetic monopoles, which are hypothetical particles that possess a single magnetic pole, either north or south. In 1931, Dirac proposed the existence of magnetic monopoles and developed a theory for their behavior. Although magnetic monopoles have not yet been observed experimentally, Dirac's work on this subject has had significant implications for the study of high-energy physics and cosmology. In conclusion, Paul Dirac made several significant contributions to quantum physics, including the development of the Dirac equation, the concept of the positron, and quantum electrodynamics. His work helped to establish the foundations of quantum mechanics and has had a significant impact on our understanding of the behavior of particles at the atomic and subatomic level. Dirac's legacy continues to inspire scientists and researchers today, and his contributions to physics remain a cornerstone of modern physics.

Richard Feynman was an American physicist who made significant contributions to the field of quantum mechanics. His work helped to establish the foundations of quantum field theory and has had a significant impact on modern physics. Feynman's most significant contribution to quantum physics was the development of the Feynman diagrams, which are a graphical representation of the mathematical formulas used to describe the behavior of particles in quantum mechanics. The Feynman diagrams helped to simplify complex calculations and make them more accessible to physicists, and they have become an essential tool for studying the behavior of particles in high-energy collisions. Feynman also made significant contributions to the understanding of the behavior of particles in condensed matter systems. In 1957, Feynman proposed the concept of the path integral formulation of quantum mechanics, which describes the behavior of particles in terms of all possible paths they could take. This concept helped to establish the foundations for the study of condensed matter physics and has become an essential tool for understanding the behavior of materials at the atomic and subatomic level. Another significant contribution of Feynman to quantum physics was his work on the interpretation of quantum mechanics. Feynman proposed the concept of the many-worlds interpretation, which states that every possible outcome of a quantum measurement exists in a separate universe. This concept helped to establish a more complete understanding of the behavior of particles in the quantum world and has had significant implications for the study of quantum information. Feynman's work also had significant implications for the development of quantum computing. In 1982, Feynman proposed the concept of the quantum computer, which is a computer that uses quantum mechanics to perform calculations. This concept helped to establish the foundations for the study of quantum computing and has become an essential tool for solving complex computational problems.

In conclusion, Richard Feynman made several significant contributions to quantum physics, including the development of the Feynman diagrams, the path integral formulation of quantum mechanics, the many-worlds interpretation, and the concept of quantum computing. His work helped to establish the foundations of quantum mechanics and has had a significant impact on our understanding of the behavior of particles at the atomic and subatomic level. Feynman's legacy continues to inspire scientists and researchers today, and his contributions to physics remain a cornerstone of modern physics.

As you can see all of these scientists made amazing and groundbreaking discoveries that will impact the way we understand

and teach quantum physics for the foreseeable future and everyone who contributed deserves credit however these were the greatest contributors to quantum theory.

Chapter 3: The Fundamental Forces

The universe is governed by four fundamental forces that dictate the behavior of all matter and energy. These forces are gravitational force, electromagnetic force, weak nuclear force, and strong nuclear force. In this chapter, we will explore each of these fundamental forces, their properties, and their importance in understanding the universe.

Gravitational Force:

The gravitational force is a fundamental force that acts between any two objects with mass. It is proportional to the mass of the objects involved and inversely proportional to the square of the distance between them. The mathematical formulation of the gravitational force is given by Newton's law of gravitation, which states that the force of attraction between two objects is equal to the product of their masses and inversely proportional to the square of the distance between them. The gravitational force is a conservative force, meaning that the work done by the force in moving an object is independent of the path taken. This property of the gravitational force makes it possible to calculate the potential energy of an object in a gravitational field and to understand the behavior of celestial bodies in orbit around each other. The gravitational force has numerous applications in physics and astronomy. It is responsible for the motion of planets around the sun and for the formation of stars and galaxies. The study of gravitational forces has led to the development of theories of gravitation, including Newton's law of gravitation and Einstein's general theory of relativity. The gravitational force also plays a crucial role in the study of cosmology and the structure of the universe. The distribution of matter and energy in the universe is affected by the gravitational force, which leads to the formation of large-scale structures such as clusters of galaxies and superclusters. Another application of the gravitational force is in the study of gravitational waves, which are ripples in the fabric of space-time that are generated by the motion of massive objects. The detection of gravitational waves has opened up a new window on the universe and has provided new insights into the behavior of black holes and other exotic objects.

Electromagnetic Force:

The electromagnetic force is a fundamental force that acts between electrically charged particles. It is a long-range force that decreases as the distance between the particles increases. The strength of the electromagnetic force is determined by the electric charge of the particles involved and their separation distance. Like the gravitational force, the electromagnetic force is also a conservative force, meaning that the work done by the force in moving an object is independent of the path taken.

The electromagnetic force has both attractive and repulsive properties, depending on the charges of the particles involved. Like charges repel each other, while opposite charges attract each other.

The mathematical formulation of the electromagnetic force is given by Coulomb's law, which states that the force between two charged particles is proportional to the product of their charges and inversely proportional to the square of the distance between them.

Electromagnetic force has numerous applications in physics and technology. It is responsible for the behavior of atoms and molecules, including the formation of chemical bonds and the absorption and emission of light. The interaction between light and matter is governed by the electromagnetic force, which is responsible for phenomena such as reflection, refraction, and diffraction. The electromagnetic force is also responsible for the functioning of electronic devices, such as computers and smartphones. The behavior of electric circuits is determined by the electromagnetic force, which allows for the transmission and manipulation of information. Another application of the electromagnetic force is in the study of electromagnetism and the development of electromagnetic theories. These theories include Maxwell's equations, which describe the behavior of electric and magnetic fields, and the theory of electromagnetism, which provides a framework for understanding the behavior of electromagnetic waves.

Weak Nuclear Force:

The weak nuclear force is a fundamental force that acts between subatomic particles, such as protons, neutrons, and electrons. It is a short-range force that operates on a scale of about 10^{-18} meters, which is much smaller than the range of the electromagnetic and gravitational forces. The weak nuclear force is a force of interaction between particles that can cause the transformation of one type of particle into another. The weak nuclear force is responsible for certain types of radioactive decay, such as beta decay, which involves the emission of an electron or a positron from a nucleus. It is also involved in other types of decay, such as alpha decay and gamma decay. The mathematical formulation of the weak nuclear force is given by the electroweak theory, which unifies the weak nuclear force with the electromagnetic force.

The weak nuclear force has numerous applications in particle physics and nuclear engineering. It is used in the study of subatomic particles and their interactions, and in the development of nuclear technologies such as nuclear power plants and nuclear weapons.

The study of the weak nuclear force has led to numerous discoveries in particle physics, including the discovery of the W and Z bosons, which are the particles responsible for mediating the weak nuclear force. The electroweak theory, which unifies the weak nuclear force with the electromagnetic force, is one of the cornerstones of the Standard Model of particle physics. Another application of the weak nuclear force is in the study of neutrinos, which are subatomic particles that are produced by the weak nuclear force. Neutrinos are difficult to detect because they interact very weakly with matter, but their study has led to numerous discoveries in particle physics and astrophysics, including the confirmation of neutrino oscillations and the measurement of the neutrino mass.

Strong Nuclear Force:

The strong nuclear force is a fundamental force that acts between subatomic particles, such as protons and neutrons. It is a short-range force that operates on a scale of about 10^{-15} meters, which is much smaller than the range of the electromagnetic and gravitational forces. The strong nuclear force is responsible for binding protons and neutrons together in the nucleus of an atom, overcoming the electromagnetic repulsion between the positively charged protons.

The strong nuclear force is one of the strongest forces in nature, with a strength that is about 100 times stronger than the electromagnetic force. It is also a force that operates independently of electric charge, meaning that it can act between particles with no net electric charge. The mathematical formulation of the strong nuclear force is given by the theory of quantum chromodynamics

(QCD), which describes the behavior of particles known as quarks and gluons.

The strong nuclear force has numerous applications in particle physics and nuclear engineering. It is used in the study of subatomic particles and their interactions, and in the development of nuclear technologies such as nuclear power plants and nuclear weapons. The study of the strong nuclear force has led to numerous discoveries in particle physics, including the discovery of the Higgs boson, which is the particle responsible for giving other particles mass. The study of the strong nuclear force has also led to the development of new technologies, such as medical imaging techniques that use radioactive isotopes. Another application of the strong nuclear force is in the study of the structure of matter. The strong nuclear force is responsible for the formation of hadrons, which are particles made up of quarks and gluons. The study of hadrons and their interactions has led to a better understanding of the structure of matter and the behavior of subatomic particles.

Importance of the Four Fundamental Forces:

The four fundamental forces are essential for understanding the behavior of matter and energy at all scales, from the subatomic level to the scale of the entire universe. They provide a framework for explaining the behavior of celestial bodies, the structure of matter, and the interactions between subatomic particles. Understanding these forces has led to numerous discoveries in physics, including the development of the standard model of particle physics and the discovery of black holes and dark matter.

The Math

The mathematical formula for calculating the gravitational force between two masses is given by Newton's law of universal gravitation:

$F = G\,(m1m2/r^2)$

where F is the gravitational force, m1 and m2 are the masses of the two objects, r is the distance between them, and G is the gravitational constant. The gravitational constant has a value of approximately 6.67×10^{-11} Nm^2/kg^2.

The formula tells us that the gravitational force between two masses is proportional to the product of their masses and inversely proportional to the square of the distance between them. This means that the force increases as the masses of the two objects increase, and decreases as the distance between them increases.

The Quantum Frontier

The mathematical formula for calculating the electromagnetic force between two charges is given by Coulomb's law:

$$F = k(q1q2/r^2)$$

where F is the electromagnetic force, q1 and q2 are the charges of the two objects, r is the distance between them, and k is Coulomb's constant. Coulomb's constant has a value of approximately $8.987 \times 10^9 \, Nm^2/C^2$.

The formula tells us that the electromagnetic force between two charges is proportional to the product of their charges and inversely proportional to the square of the distance between them. This means that the force increases as the charges of the two objects increase, and decreases as the distance between them increases.

The mathematical formula for calculating weak nuclear force is given by Fermi's interaction:

$$F = G\_F^2 \, (cos\theta\_C)^2 \, (m1m2)/(r^2)$$

where F is the weak nuclear force, G_F is the Fermi coupling constant, θ_C is the Cabibbo angle, m1 and m2 are the masses of the two particles, and r is the distance between them.

The formula tells us that the weak nuclear force is proportional to the square of the Fermi coupling constant and the square of the Cabibbo angle, which are both fundamental constants of nature. It is also proportional to the product of the masses of the particles and inversely proportional to the square of the distance between them

The mathematical formula for calculating the strong nuclear force is given by the quantum chromodynamics (QCD) theory:

$F\_s = - \alpha\_s (4/3) (\pi r^2) (\rho\_1 \rho\_2) (1 - e^{(-m r)})/r^2$

where F_s is the strong nuclear force, α_s is the strong coupling constant, r is the distance between the particles, ρ_1 and ρ_2 are the quark densities, and m is the mass of the exchanged gluon.

The formula tells us that the strong nuclear force is proportional to the strong coupling constant, the quark densities, and the area of the interaction. It is also inversely proportional to the square of the distance between the particles and the mass of the exchanged gluon. The exponential term in the formula accounts for the confinement of quarks within the nucleus.

Chapter 4: The Fundamental particles

Elementary particle physics is a branch of physics that deals with the study of the fundamental particles and forces that make up the universe. These particles include leptons, quarks, bosons, and other exotic particles. The study of these particles and their interactions is essential in understanding the underlying nature of the universe. In this paper, we will discuss the fundamental particles and forces, their properties, interactions, and the experimental techniques used to study them.

The fundamental particles are the building blocks of matter, and they are divided into two categories: fermions and bosons. Fermions are the particles that make up matter, while bosons are the force-carrying particles.

Fermions are particles with half-integer spin, and they obey the Pauli exclusion principle, which states that no two fermions can occupy the same quantum state simultaneously. There are two types of fermions: quarks and leptons.

Intrinsic angular momentum, also known as spin, is one of the fundamental properties of particles in the universe. It is a quantum mechanical property of particles that manifests as an angular momentum even when the particle is not rotating or moving. In this paper, we will explore the concept of intrinsic angular momentum, its properties, and its applications in physics.

Intrinsic angular momentum is a fundamental property of particles that cannot be explained by classical mechanics. The spin of a particle is a quantum mechanical property, meaning that it can only be described by the principles of quantum mechanics. Spin is a vector quantity, meaning that it has both magnitude and direction. The magnitude of the spin is given by the spin quantum number, which is a half-integer value. For example, the spin of an electron is 1/2, while the spin of a proton is 1/2. One of the most unique properties of intrinsic angular momentum is that it is always quantized. This means that the spin of a particle can only take on certain discrete values, rather than any value in between. For example, the spin of an electron can only be measured as +1/2 or -1/2. This quantization of spin is a consequence of the principles of quantum mechanics and has been experimentally verified. Another property of intrinsic angular momentum is that it is conserved. This means that the total spin of a system of particles remains constant, even as the particles interact with each other. This conservation of spin is a consequence of the laws of physics, which dictate that the total angular momentum of a closed system must be conserved. Intrinsic angular momentum has numerous applications in physics, including in the study of atomic and subatomic particles. The spin of electrons is one of the key properties that determines the behavior of atoms and molecules. For example, the spin of electrons is responsible for the magnetic properties of materials, as well as the splitting of spectral lines in atomic spectra.

Intrinsic angular momentum also plays a crucial role in the study of subatomic particles, such as protons and neutrons. The spin of these particles determines their magnetic properties, as well as their interactions with other particles. The study of subatomic particles and their spin properties has led to numerous discoveries in physics, including the development of the standard model of particle physics.

Fermi-Dirac statistics, also known as Fermi statistics, are a type of quantum statistics that describe the behavior of a large collection of identical particles that obey the Pauli exclusion principle. These particles are called fermions, and they include electrons, protons, and neutrons.

The Fermi-Dirac statistics were first introduced in 1926 by Enrico Fermi and Paul Dirac, who applied the statistics to describe the behavior of electrons in a solid. This led to the development of the Fermi gas model, which describes the behavior of a collection of non-interacting, spin-half fermions in a box.

One of the key features of Fermi-Dirac statistics is that they allow for a maximum of one fermion to occupy a given quantum state. This is due to the Pauli exclusion principle, which states that no two fermions can occupy the same quantum state simultaneously. As a result, the energy levels of a Fermi gas are filled in a stepwise manner, with each energy level occupied by only one fermion.

Another important property of Fermi-Dirac statistics is that they lead to the phenomenon of degeneracy pressure, which arises when fermions are compressed into a small volume. This pressure arises from the fact that, as the fermions are compressed, their energy levels become more closely spaced, leading to a larger number of occupied states at the lowest energies. This increased occupancy of low-energy states leads to an increase in the pressure, which can oppose the force of gravity and prevent the collapse of a star or other massive object.

Fermi-Dirac statistics also play an important role in a wide range of other fields, including solid-state physics, nuclear physics, and astrophysics. In solid-state physics, the statistics are used to describe the behavior of electrons in a crystal lattice, and they provide a theoretical framework for understanding phenomena such as electrical conductivity and magnetism. In nuclear physics, the statistics are used to describe the behavior of protons and neutrons in a nucleus, and they help to explain the properties of stable and unstable isotopes. In astrophysics, the statistics are used to describe the behavior of matter in the cores of stars, and they play a crucial role in understanding the behavior of white dwarfs and neutron stars.

In conclusion, Fermi-Dirac statistics are an important tool for understanding the behavior of a large collection of identical fermions, such as electrons, protons, and neutrons. These statistics describe the behavior of fermions under conditions where the Pauli exclusion principle applies, and they are important for understanding a wide range of phenomena in fields such as solid-state physics, nuclear physics, and astrophysics.

Up Quarks

Up quarks are one of the six types of quarks, which are elementary particles that make up protons and neutrons, the building blocks of atomic nuclei. Up quarks, along with down quarks, are the lightest of the six quarks and are the most common in the universe. They are characterized by their charge, spin, and flavor quantum numbers, which describe their fundamental properties. Up quarks have a positive charge of +2/3 e, where e is the elementary charge. This means that up quarks interact electromagnetically and are repelled by other particles with a positive charge, like other up quarks or protons. Up quarks also have a half-integer spin of 1/2, which is a fundamental quantum property that describes their intrinsic angular momentum. Up quarks interact through the strong nuclear force, which is responsible for binding quarks together to form protons and neutrons. The strong force is mediated by gluons, which are themselves elementary particles. Up quarks also interact through the weak nuclear force, which is responsible for some forms of radioactive decay and neutrino interactions. In addition to their charge and spin, up quarks also have a flavor quantum number, which distinguishes them from other types of quarks. Up quarks have an up flavor, while the other five types of quarks have a down, charm, strange, top, or bottom flavor. These flavors are not related to the taste of the particles, but rather describe their different properties. Up quarks are studied in particle physics experiments that aim to understand the fundamental building blocks of matter and the interactions between them. These experiments involve accelerating particles to high energies and colliding them together, creating a shower of other particles that can be detected and analyzed. Up quarks are also important in cosmology, the study of the universe as a whole. The abundance of up quarks and other particles in the early universe played a crucial role in determining the distribution of matter in the universe today. Up quarks are an essential component of the universe, making up a significant fraction of the protons and neutrons that form atomic nuclei. Their charge, spin, and flavor quantum numbers distinguish them from other types of quarks and allow them to interact through the strong and weak nuclear forces. The study of up quarks and other fundamental particles is a crucial part of understanding the universe at its most fundamental level.

Down Quarks

Down quarks have a negative charge of -1/3 e, where e is the elementary charge. This means that down quarks interact electromagnetically and are attracted to other particles with a positive charge, like up quarks or protons. Down quarks also have a half-integer spin of ½. In addition to their charge and spin, down quarks also have a flavor quantum number, which distinguishes them from other types of quarks. Down quarks have a down flavor. Down quarks interact through the strong nuclear force, which is responsible for binding quarks together to form protons and neutrons. The strong force is mediated by gluons, which are themselves elementary particles. Down quarks also interact through the weak nuclear force, which is responsible for some forms of radioactive decay and neutrino interactions. Down quarks are studied in particle physics experiments that aim to understand the fundamental building blocks of matter and the interactions between them. These experiments involve accelerating particles to high energies and colliding them together, creating a shower of other particles that can be detected and analyzed.

Charm Quarks

Charm quarks have a charge of +2/3 e, where e is the elementary charge. This means that they interact electromagnetically and are attracted to particles with a negative charge, such as down quarks or electrons. Charm quarks also have a half-integer spin of ½. In addition to their charge and spin, charm quarks also have a flavor quantum number, which distinguishes them from other types of quarks. Charm quarks have a charm flavor. Charm quarks interact through the strong nuclear force, which is responsible for binding quarks together to form protons and neutrons. Charm quarks also interact through the weak nuclear force, which is responsible for some forms of radioactive decay and neutrino interactions. Charm quarks were first discovered in 1974 by a team of physicists at the Stanford Linear Accelerator Center (SLAC) in California, USA. The discovery of charm quarks was an important milestone in particle physics research, as it confirmed the existence of a third generation of quarks beyond the up, down, and strange quarks that were known at the time. Charm quarks are important in particle physics research because they can be used to study the properties of other particles. For example, charm quarks can be produced in collisions between protons and antiprotons at the Large Hadron Collider (LHC) in Switzerland, and the properties of the resulting particles can be studied to gain insights into the behavior of other particles. In addition to their role in particle physics research, charm quarks are also important in the study of the early universe. The abundance of charm quarks and other particles in the early universe played a crucial role in determining the distribution of matter in the universe today.

Strange Quarks

Like all quarks, strange quarks have an electric charge, but they are negatively charged. They also have a property called strangeness, which is a quantum number that describes their tendency to interact strongly with other strange particles. Strange quarks have a strangeness of -1, meaning they have one unit of strangeness. Strange quarks have a mass of approximately 80 to 130 MeV/c^2, which is much greater than the mass of up and down quarks. This makes strange quarks much more difficult to produce in experiments. Strange quarks can also combine with other quarks to form hadrons, which are particles made up of quarks held together by the strong nuclear force. Mesons, which are hadrons made up of a quark and an antiquark, can contain strange quarks. One of the most interesting properties of strange quarks is their ability to decay through the weak nuclear force. This occurs through a process called strangeness-changing weak decay, where a strange quark changes into a down quark or an up quark, and emits a W boson. This decay process has been observed in a variety of particles, including K mesons and lambda baryons. Strange quarks also play a role in the study of particle physics through their interactions with other particles. They can interact with up and down quarks through the strong nuclear force to form hadrons, and with other strange quarks through the strong and weak nuclear forces to form multi-strange hadrons. These interactions can provide insight into the properties of the strong nuclear force, and into the behavior of particles in extreme conditions, such as those found in the early universe or in neutron stars. Strange quarks are important in the study of particle physics because they provide insight into the properties of the strong nuclear force and into the behavior of particles under extreme conditions. They also play a role in the formation of hadrons, which are the building blocks of matter. In addition, the decay of strange particles through the weak nuclear force can provide important information about the behavior of subatomic particles .Strange quarks are an important component of the Standard Model of particle physics, providing insight into the behavior of particles under extreme conditions and into the properties of the strong nuclear force. Their unique ability to decay through the weak nuclear force has been observed in a variety of particles, and their interactions with other particles provide important information about the behavior of matter at a fundamental level

Top Quarks

The top quark is the heaviest known elementary particle, with a mass of approximately 173 GeV/c^2. It was discovered in 1995 by the CDF and D0 collaborations at the Fermi National Accelerator Laboratory (Fermilab) in Illinois, USA. Since its discovery, the top quark has played a crucial role in testing the standard model of particle physics and searching for new physics beyond the standard model. In this paper, we will discuss the properties and interactions of the top quark, as well as its discovery and current status in particle physics research. The top quark is a member of the third generation of quarks, along with the bottom and charm quarks. It has an electric charge of +2/3 e and a spin of 1/2. Its mass is approximately 173 GeV/c^2, which is about 35 times the mass of the next heaviest quark, the bottom quark. The top quark is a very short-lived particle with a lifetime of approximately 5 × 10^-25 seconds. It decays almost exclusively to a W boson and a bottom quark. This is due to the fact that the top quark is the only quark heavy enough to decay to a W boson and another quark. The W boson then decays into a charged lepton (electron, muon, or tau) and a neutrino or into two quarks.The top quark interacts primarily through the strong force and the weak force. The strong force is responsible for binding quarks together to form hadrons, such as protons and neutrons. The weak force is responsible for the decay of the top quark into a W boson and a bottom quark.The strong force is mediated by gluons, which are massless particles that carry a color charge. Quarks also carry a color charge, which comes in three varieties: red, green, and blue. The strong force is so strong that it is impossible to observe a single quark in isolation; instead, quarks are always found bound together in hadrons.The weak force is responsible for the decay of the top quark. The weak force is mediated by the W and Z bosons, which are massive particles. The W boson carries a charge of +1 or -1, while the Z boson is electrically neutral. The weak force is much weaker than the strong force, which is why the top quark has a relatively long lifetime compared to other elementary particles.The top quark was discovered in 1995 by the CDF and D0 collaborations at Fermilab. This discovery was the culmination of years of research and development of new accelerator and detector technology. The top quark was discovered through its production in high-energy proton-antiproton collisions at Fermilab's Tevatron accelerator. The Tevatron was capable of colliding protons and antiprotons at energies of up to 2 TeV, which is approximately 20 times the mass of the top quark. The CDF and D0 collaborations searched for the top quark by looking for the distinctive decay signature of a top quark pair. When a top quark pair is produced, each top quark decays to a W boson and a bottom quark. The W bosons then decay into charged leptons and neutrinos or into quark-antiquark pairs. By looking for events with the appropriate number and type of particles in the detector, the collaborations were able to identify the top quark signals.

Bottom Quark

Bottom quark, also known as a beauty quark, is a type of elementary particle in the standard model of particle physics. It is the third-heaviest of all quarks, with a mass of approximately 4.2 GeV/c^2, and has an electric charge of -1/3. Like other quarks, it interacts via strong, weak, and electromagnetic forces, and can form bound states with other quarks to create composite particles such as mesons and baryons.

The existence of the bottom quark was first proposed in the 1970s by theorists Sheldon Glashow, John Iliopoulos, and Luciano Maiani to explain certain experimental observations. In 1977, the bottom quark was finally discovered independently by two research groups, one at the Fermi National Accelerator Laboratory (Fermilab) in the United States and the other at the European Organization for Nuclear Research (CERN) in Switzerland. The Fermilab experiment used a particle accelerator to produce high-energy proton-antiproton collisions, which were then detected by a large detector known as the Tevatron. The CERN experiment used a similar technique, colliding protons with antiprotons and detecting the resulting particles using a detector called the Big European Bubble Chamber.

As mentioned earlier, the bottom quark has a mass of approximately 4.2 GeV/c^2, which is around four times heavier than the up quark and two times heavier than the charm quark. Its electric charge is $-1/3$, and it has a spin of $1/2$. It is also a member of the third generation of quarks, which includes the top quark.

Like other quarks, the bottom quark interacts via the strong, weak, and electromagnetic forces. The strong force is responsible for binding quarks together to form composite particles, such as protons and neutrons. The weak force is responsible for certain types of particle decay, such as beta decay. The electromagnetic force is responsible for interactions between charged particles. The bottom quark can form bound states with other quarks to create composite particles such as mesons and baryons. For example, a meson consisting of a bottom quark and its anti-particle, the anti-bottom quark, is known as a B meson. B mesons are particularly interesting because they can decay in a variety of ways, including the rare decay into two muons, which has been used to search for new physics beyond the standard model.

The properties of the bottom quark and its bound states, such as B mesons, have been extensively studied by experimental physicists over the past few decades. These studies have contributed significantly to our understanding of the standard model of particle physics and have helped to test its predictions. In particular, the discovery of CP violation in the decay of B mesons, which was first observed by the Belle and BaBar experiments in the early 2000s, was a major breakthrough that provided crucial evidence for the standard model's explanation of why there is more matter than antimatter in the universe. In addition to fundamental research, the properties of the bottom quark have also found practical applications in areas such as medical imaging. For example, a technique known as positron emission tomography (PET) uses radioactive isotopes that decay into particles that include B mesons. By detecting these particles, doctors can create detailed images of the body that can help diagnose and treat diseases.

Electron Leptons

Electrons are elementary particles with a negative electric charge of -1. They are part of the lepton family and belong to the first generation of elementary particles. Electrons have a mass of $9.10938356 \times 10^{-31}$ kilograms, making them one of the lightest known particles. Electrons are fundamental particles, meaning they cannot be broken down into smaller components. They are also point particles, meaning they have no spatial extent. Electrons are involved in many of the fundamental processes of the universe, including chemical bonding, electricity, and magnetism. The behavior of electrons is described by the laws of quantum mechanics, which govern the behavior of particles at the atomic and subatomic levels. One of the most important features of electrons is their wave-particle duality. This means that electrons can behave like both waves and particles, depending on how they are observed.

In some experiments, electrons behave like particles, while in others, they behave like waves. This duality is a fundamental aspect of quantum mechanics and is crucial to our understanding of the behavior of electrons.Electrons also have a property called spin, which is a form of intrinsic angular momentum. Spin is a quantum mechanical property that cannot be explained by classical physics. Spin is measured in units of $h/2\pi$, where h is the Planck constant. Electrons have a spin of 1/2, which means they can exist in two possible states, up or down. The behavior of electrons is also influenced by their interaction with other particles. Electrons can be scattered or absorbed by other particles, and their paths can be influenced by electric and magnetic fields. Electrons can also be excited to higher energy states, which can lead to the emission of light or the ionization of atoms.

The properties and behavior of electrons have many practical applications. One of the most important applications is in electronics. Electrons are the charge carriers in electrical circuits and are responsible for the operation of electronic devices such as computers, smartphones, and televisions. The properties of electrons also play a crucial role in the design and fabrication of semiconductor devices, which are used in a wide range of electronic applications. Electrons are also important in the field of chemistry. The behavior of electrons in atoms and molecules is responsible for the formation of chemical bonds and the properties of compounds. The study of electron behavior in chemical systems is known as quantum chemistry and is used to predict the properties of new materials and chemical reactions. In the field of medicine, electrons are used in a variety of imaging techniques, such as X-ray and computed tomography (CT) scans. These imaging techniques use the interaction of electrons with matter to produce images of the internal structures of the body.

Muons

Muons are elementary particles that belong to the lepton family. They are similar to electrons and have the same negative charges but are much more massive, with a mass of approximately 207 times that of an electron. Muons were discovered by Carl D. Anderson in 1936 while studying cosmic rays. Muons are unstable particles with a half-life of approximately 2.2 microseconds. They decay into an electron, two neutrinos, and an antineutrino. Muons are created when high-energy cosmic rays collide with particles in the upper atmosphere, producing a shower of particles that includes muons. Like electrons, muons have a spin of 1/2 and obey the laws of quantum mechanics. They are subject to the same electromagnetic forces as electrons and can be accelerated by electric and magnetic fields. In addition, muons interact weakly with other particles, allowing them to participate in processes such as beta decay. Muons have been used in a variety of scientific experiments, including the measurement of the muon's magnetic moment and the discovery of the muon neutrino. Muons have also been used to study the properties of materials and to detect hidden chambers in ancient Egyptian pyramids. One of the most famous experiments involving muons is the "muon g-2" experiment. This experiment measures the magnetic moment of the muon, which is a measure of its intrinsic spin. The magnetic moment of the muon is influenced by the surrounding magnetic field, and the "g-factor" is a measure of the degree to which the muon's spin interacts with the magnetic field. The results of the muon g-2 experiment have provided important constraints on the Standard Model of particle physics. Muons have also been used in experiments to study the fundamental forces of nature. For example, muons have been used to study the weak force, which is responsible for radioactive decay. By measuring the rate of beta decay of muons, physicists can determine the strength of the weak force. In conclusion, muons are an important class of elementary particles that have been used in a wide variety of scientific experiments. They are similar to electrons but much more massive and unstable. Muons are subject to the same electromagnetic forces as electrons and can be accelerated by electric and magnetic fields. They also interact weakly with other particles, allowing them to participate in processes such as beta decay. Muons have provided important insights into the fundamental forces of nature and continue to be an important tool for scientists studying the subatomic world.

Taus

Tau particles, also known as tau leptons, are one of the six known types of leptons in particle physics. They were first discovered in 1975 at the Stanford Linear Accelerator Center (SLAC) in California. Tau particles are similar to electrons and muons, but they are much heavier, with a mass of approximately 1.78 GeV/c², which is roughly 3,500 times heavier than an electron. Like other leptons, tau particles are elementary particles, which means they are not composed of smaller particles. They have a negative electric charge, and their spin is 1/2. Tau particles are produced in high-energy collisions, such as those that occur in particle accelerators or in cosmic rays. Tau particles are unstable and decay very quickly, with a mean lifetime of about 2.9×10^{-13} seconds. When a tau particle decays, it produces a variety of particles, including neutrinos, which are electrically neutral and interact very weakly with matter. Because neutrinos are difficult to detect, their production in tau decay provides an important way to study these elusive particles. The study of tau particles and their decay properties has played an important role in testing the predictions of the Standard Model of particle physics. In particular, the observation of tau neutrinos provided strong evidence for the existence of a third generation of particles, which was predicted by the Standard Model. The discovery of the tau particle also led to the development of new experimental techniques and technologies, such as the use of large detectors to study high-energy particle collisions.

In addition to their role in fundamental particle physics research, tau particles also have practical applications in medical imaging. Positron emission tomography (PET) is a medical imaging technique that uses the decay of positrons, which are antiparticles of electrons, to produce images of the body. The decay of a positron produces two gamma rays, which can be detected by a PET scanner.

One way to produce positrons is to use a cyclotron to accelerate protons to high energies and then use them to produce isotopes that decay by positron emission. One such isotope is fluorine-18, which is produced by bombarding oxygen-18 with protons. When fluorine-18 decays by positron emission, it produces a positron and a neutrino. The positron quickly collides with an electron in the body and annihilates, producing two gamma rays that are detected by the PET scanner. However, about 70% of the time, the positron produced in fluorine-18 decay interacts with an electron in the body and produces a tau particle. The tau particle quickly decays, producing additional gamma rays that are detected by the PET scanner. This process, known as positron annihilation with prompt gamma emission (PAPGE), can provide additional information about the distribution of the radiotracer in the body, leading to improved image quality and diagnostic accuracy. In conclusion, tau particles are important elementary particles that have played a key role in the development of modern particle physics. They have helped to test the predictions of the Standard Model and have led to the development of new experimental techniques and technologies. In addition, the decay properties of tau particles have practical applications in medical imaging, particularly in the development of new PET imaging techniques.

Electron Neutrino

The electron neutrino is denoted by the symbol ν_e and is the lightest of the three neutrino flavors. It is created in various nuclear reactions, such as beta decay, electron capture, and nuclear fusion.

The electron neutrino has no electric charge and a very small mass, which is currently estimated to be less than 1 eV/c². This makes the electron neutrino one of the least massive particles known to exist. The electron neutrino has a half-integer spin of 1/2 and is a fermion, which means it follows the Fermi-Dirac statistics. It also has a corresponding antiparticle, the electron antineutrino ($\overline{\nu}_e$), which has the opposite quantum numbers. The electron neutrino interacts only weakly with matter and is therefore very difficult to detect. It interacts via the weak force, which is mediated by the exchange of W and Z bosons. The most common interaction of the electron neutrino is the elastic scattering off electrons, which is known as the neutral current interaction. In this process, the neutrino exchanges a Z boson with the electron and changes its direction. Another important interaction of the electron neutrino is the charged current interaction, which occurs when the neutrino interacts with a charged lepton, such as an electron. In this process, the neutrino emits a W boson, which is absorbed by the lepton, causing it to change into a different flavor. For example, an electron neutrino can interact with an electron and produce a muon neutrino, which can then interact with another electron and produce a tau neutrino.

Due to its weak interaction with matter, the electron neutrino is very difficult to detect. However, there are several methods that have been developed for detecting neutrinos, such as the radiochemical detection method, the Cherenkov radiation detection method, and the scintillation detection method. The radiochemical detection method involves exposing a large quantity of a certain element, such as chlorine or gallium, to a source of neutrinos, such as the Sun. If a neutrino interacts with an atom in the detector, it will produce a radioactive isotope, which can be detected and measured. The Cherenkov radiation detection method involves detecting the faint flashes of light that are emitted when a charged particle, such as an electron, moves through a transparent material at a speed greater than the speed of light in that material. This method has been used to detect neutrinos produced in cosmic ray showers and in nuclear reactors. The scintillation detection method involves using a material that emits light when it is struck by a charged particle, such as an electron. When a neutrino interacts with the material, it produces a charged particle, which in turn produces light that can be detected and measured.

Muon Neutrinos

Muon neutrinos are elementary particles that belong to the lepton family, which includes electrons, muons, and taus. Neutrinos are elusive particles that have no electric charge, and their masses are much smaller than those of other particles. They interact weakly with matter and are capable of passing through vast amounts of matter without being affected. They are also produced in large numbers by natural processes, such as nuclear reactions in the sun and in cosmic ray interactions with the Earth's atmosphere. Muon neutrinos are specifically created in the decay of muons, which are heavier and more unstable cousins of electrons. Muons are produced naturally in cosmic ray interactions with the atmosphere and in the decay of other particles. They are also produced artificially in particle accelerators, where they can be studied in detail. The muon neutrino was first postulated by physicists in the mid-20th century, as a way to explain certain experimental results that were inconsistent with known physics. In the 1960s and 1970s, experiments began to confirm the existence of muon neutrinos, and by the 1980s, they had been detected in several experiments. The detection of muon neutrinos is a challenging process, due to their weak interactions with matter. Typically, detectors must be large and sensitive to detect the rare interactions that occur between neutrinos and other particles. The most common method for detecting muon neutrinos is to observe the muons produced when the neutrinos interact with matter. For example, in a detector filled with water or ice, a muon neutrino may interact with an atomic nucleus, producing a muon that can be detected as it travels through the medium. The study of muon neutrinos is important for several reasons. First, they provide valuable information about the properties of neutrinos in general, which are still not well understood. Second, they can be used to study the properties of the weak nuclear force, which is responsible for the interactions between particles that produce muon neutrinos. Finally, they can be used to study astrophysical processes, such as supernovae, which produce large numbers of neutrinos. In recent years, several experiments have been conducted to study muon neutrinos in more detail. One such experiment is the MINOS experiment, which is located at Fermilab in the United States. This experiment uses a neutrino beam produced by a particle accelerator to study the properties of muon neutrinos as they travel through the Earth. Another experiment is the IceCube experiment, which is located at the South Pole. This experiment uses a large detector made of ice to study neutrinos produced by astrophysical processes, including muon neutrinos.

Tau Neutrino

The tau neutrino, like the electron and muon neutrinos, is a spin-1/2 particle with no electric charge. It interacts only weakly with matter, passing through it almost completely undetected. The tau neutrino is much more massive than the electron and muon neutrinos, with a mass of about 17 MeV/c^2, compared to less than 1 eV/c^2 for the other two flavors. This makes the tau neutrino much harder to produce and detect, and it was not observed directly until 2000, by the DONUT collaboration at Fermilab.

The detection of tau neutrinos requires high-energy particle accelerators and sophisticated detectors. In general, neutrinos are detected by their interactions with matter, which produce secondary particles that can be detected. However, the weak interaction of tau neutrinos makes them difficult to detect, as they rarely interact with matter. One technique for detecting tau neutrinos is to produce them in high-energy collisions and look for the characteristic decay products of the tau lepton, which has a very short lifetime of about 10^-13 seconds.

Tau neutrinos have a number of important applications in astrophysics and particle physics. In astrophysics, they are produced in the interactions of cosmic rays with the Earth's atmosphere, and can be used to study the properties of these high-energy particles. In particle physics, they are produced in high-energy colliders and used to study the properties of the weak force and the Higgs boson. They may also play a role in theories of physics beyond the standard model, such as supersymmetry and extra dimensions.

Gluons

Gluons are elementary particles that are responsible for mediating the strong nuclear force, one of the four fundamental forces in nature. They play a crucial role in the structure of matter, binding quarks together to form protons, neutrons, and other particles. In this paper, we will discuss the properties of gluons, their interactions with other particles, and their role in the strong nuclear force. Gluons are massless particles, which means that they travel at the speed of light. They also carry a color charge, which is the property that allows them to interact with other particles through the strong nuclear force. There are eight different types of gluons, each with a different combination of color and anti-color charge. These gluons can interact with quarks and other gluons through the exchange of virtual gluons.

Gluons interact with other particles through the strong nuclear force, which is responsible for binding quarks together to form protons, neutrons, and other particles. When two quarks come close to each other, they exchange gluons, which transfer the strong nuclear force between them. This interaction is what binds quarks together to form composite particles. Gluons also interact with other gluons through the exchange of virtual gluons. This interaction can lead to the production of new particles, such as mesons or baryons, which are composed of quarks and anti-quarks or three quarks, respectively.

The strong nuclear force is responsible for holding atomic nuclei together. It is the strongest of the four fundamental forces and is mediated by gluons. When two or more quarks come close to each other, they exchange gluons, which transfer the strong nuclear force between them. This interaction is what binds quarks together to form composite particles, such as protons and neutrons. The strong nuclear force is also responsible for the stability of atomic nuclei. Without the strong nuclear force, atomic nuclei would not be able to overcome the electrostatic repulsion between the positively charged protons in the nucleus.

In conclusion, gluons are elementary particles that play a crucial role in the structure of matter. They are massless particles that carry a color charge, which allows them to interact with other particles through the strong nuclear force. Gluons interact with quarks and other gluons through the exchange of virtual gluons, and this interaction is what binds quarks together to form composite particles. The strong nuclear force is responsible for holding atomic nuclei together and is mediated by gluons.

Photons

The photon is a fundamental particle in the universe, and it is the mediator of the electromagnetic force. Photons are massless and travel at the speed of light. They can behave both as particles and waves, a phenomenon known as wave-particle duality. This paper will discuss the properties and behaviors of photons, as well as their various applications in technology and research. Photons are elementary particles that have zero rest mass and travel at the speed of light. They are electrically neutral, meaning they are not affected by electric fields. However, they are affected by magnetic fields and can produce their own magnetic field when they are in motion. Photons are also spin-1 particles, meaning they have an intrinsic angular momentum. The energy of a photon is directly proportional to its frequency, which is related to its wavelength through the speed of light. This relationship is given by the equation $E = hf$, where E is the energy of the photon, h is Planck's constant, and f is the frequency of the photon. The energy of a single photon is relatively small, but large numbers of photons can produce significant amounts of energy.

Photons can exhibit both wave-like and particle-like behavior, a phenomenon known as wave-particle duality. This duality is a fundamental concept of quantum mechanics and describes the dual nature of matter and energy. In wave-like behavior, photons can interfere with each other, producing patterns of constructive and destructive interference. In particle-like behavior, photons can be detected as discrete particles, such as when they interact with a detector in a photon-counting experiment.

Another property of photons is polarization, which describes the orientation of the electric and magnetic fields that make up the photon's electromagnetic wave. Polarization is important in many applications, including telecommunications and imaging technologies.

Photons have a wide range of applications in technology and research. In telecommunications, they are used to transmit information through fiber optic cables. In medicine, they are used in imaging technologies such as X-rays, CT scans, and MRI scans. In manufacturing, they are used in lasers for cutting and welding materials. Research on photons has also led to important discoveries in the field of quantum mechanics, such as the concept of entanglement. Entanglement occurs when two or more particles become correlated in a way that their states are dependent on each other, even when separated by large distances. This phenomenon has important implications for quantum computing and communication.

Photons are fundamental particles that play a crucial role in the universe as mediators of the electromagnetic force. They exhibit wave-particle duality and have properties such as polarization and energy that make them useful in a variety of applications. Research on photons has led to important discoveries in quantum mechanics and has opened up new possibilities for technology and communication.

Higgs

The Higgs particle, also known as the Higgs boson, is a fundamental particle in the Standard Model of particle physics. It was first proposed in the 1960s by physicist Peter Higgs, along with several other physicists, as a way to explain why particles have mass. The existence of the Higgs particle was confirmed in 2012 by experiments at the Large Hadron Collider (LHC) at CERN in Geneva, Switzerland. The Higgs particle is unique among the fundamental particles in that it is associated with a field, the Higgs field, which is present throughout the universe. The Higgs field is a type of quantum field, similar to the electromagnetic field, and interacts with other particles in a way that gives them mass. The Higgs field is a scalar field, meaning that it has a single value at each point in space. When particles interact with the Higgs field, they experience a resistance, which gives them mass. The strength of the interaction between the particle and the Higgs field determines the mass of the particle. The Higgs particle itself is a type of boson, which means that it carries a force between particles. Unlike other bosons, however, the Higgs particle does not carry a force that acts at long distances. Instead, it interacts with other particles only when they are very close together, within a few femtometers.

The discovery of the Higgs particle was a major milestone in the history of particle physics. It confirmed the existence of the Higgs field, which had been theorized for decades, and provided a key piece of evidence supporting the Standard Model of particle physics. The Standard Model describes the interactions of all known fundamental particles, including the Higgs particle, and is one of the most successful theories in physics. However, the discovery of the Higgs particle also raised new questions and challenges for particle physics. For example, the mass of the Higgs particle is much larger than would be expected based on the other particles in the Standard Model, leading some physicists to speculate that there may be new physics beyond the Standard Model. In addition, the Higgs field is related to the process of electroweak symmetry breaking, which is believed to be responsible for the asymmetry between matter and antimatter in the universe.

In conclusion, the Higgs particle is a fundamental particle in the Standard Model of particle physics that is associated with a quantum field, the Higgs field, which gives other particles mass. Its discovery in 2012 was a major milestone in the history of particle physics, confirming the existence of the Higgs field and providing key evidence supporting the Standard Model. However, it also raised new questions and challenges for particle physics, and its properties and interactions continue to be studied by physicists around the world.

Z boson

Nicholas Tassy-Becz

The Z boson, also known as the Z boson or the intermediate vector boson Z, is a subatomic particle that is an important component of the Standard Model of particle physics. Like the W bosons, the Z boson is responsible for mediating the weak nuclear force, which is one of the four fundamental forces of nature. The discovery of the Z boson, along with the W bosons, provided important confirmation of the electroweak theory, which unifies the weak nuclear force with electromagnetism. The Z boson was first observed in 1983 at the European Organization for Nuclear Research (CERN) in Switzerland, using the Super Proton Synchrotron (SPS) particle accelerator. The discovery was the result of a collaboration between CERN and the Institute for Nuclear Research at Dubna in Russia, and was announced in a joint paper in July of that year. The experiment involved colliding protons with antiprotons and analyzing the resulting particles produced in the collision. The Z boson has a mass of approximately 91.2 GeV/c^2, which is about 90 times the mass of the proton. It is electrically neutral and has a spin of 1. The Z boson is responsible for mediating the weak nuclear force, which is responsible for processes such as beta decay, which occurs when a neutron decays into a proton, an electron, and a neutrino. The weak force is also involved in processes that occur in the sun, such as nuclear fusion. Like the W bosons, the Z boson can be produced in particle accelerators by colliding particles such as protons and antiprotons. The Z boson can also be produced in natural processes, such as in the decay of certain radioactive isotopes. The decay of the Z boson can produce a range of particles, including leptons (such as electrons, muons, and tau particles) and quarks (the building blocks of protons and neutrons). The discovery of the Z boson was an important confirmation of the electroweak theory, which unifies the weak nuclear force with electromagnetism. The theory was proposed by Sheldon Glashow, Abdus Salam, and Steven Weinberg in the 1970s, and was later experimentally confirmed with the discovery of the W and Z bosons. The electroweak theory is an important component of the Standard Model of particle physics, which describes the fundamental particles and forces of nature. In addition to its importance in confirming the electroweak theory, the Z boson has also been used in a range of experimental studies in particle physics. For example, experiments at CERN have used the Z boson to study the properties of the weak nuclear force, to search for new particles, and to investigate the structure of the proton.

In conclusion, the Z boson is an important subatomic particle that is responsible for mediating the weak nuclear force. Its discovery in 1983 provided important confirmation of the electroweak theory, and it continues to be an important tool in experimental studies in particle physics.

W boson

The W boson is one of the elementary particles that mediate the weak nuclear force, which is responsible for radioactive decay and nuclear fusion in stars. It was first proposed by Sheldon Glashow, Abdus Salam, and Steven Weinberg in the 1970s as part of the electroweak theory. The W boson is a gauge boson, which means it is responsible for the exchange of energy and momentum between particles. In this paper, we will explore the properties, interactions, and significance of the W boson in particle physics.

The W boson has a mass of approximately 80 GeV/c^2 and a spin of 1. It comes in two varieties, the positively charged W+ boson and the negatively charged W- boson. They are antiparticles of each other and can be produced in pairs via high-energy collisions. The W boson also has a very short lifetime, with a mean decay time of approximately 10^-25 seconds.

The W boson is responsible for the weak nuclear force, which is one of the four fundamental forces in nature, along with the electromagnetic, strong nuclear, and gravitational forces. The weak force is responsible for radioactive decay, where a nucleus emits a W boson to transform into a different element. The W boson is also involved in the fusion of hydrogen nuclei in stars, where two protons combine to form a helium nucleus and release energy in the form of light and heat. The W boson can interact with both leptons and quarks. Leptons are elementary particles that do not interact via the strong nuclear force, such as electrons, muons, and tau particles. Quarks are the building blocks of protons and neutrons, which make up the nucleus of an atom. The W boson can change the flavor of quarks by converting one type of quark into another type. This is known as weak flavor mixing or CKM mixing, named after the physicists Cabibbo, Kobayashi, and Maskawa who first proposed it.

The discovery of the W boson was a significant achievement in particle physics, as it provided experimental evidence for the electroweak theory. This theory unifies the electromagnetic and weak nuclear forces into a single electroweak force, which is mediated by the W and Z bosons. The W boson has also played a crucial role in the study of neutrinos, which are neutral particles that only interact via the weak nuclear force. Neutrinos are produced in the sun and in cosmic rays and can pass through matter without interacting, making them difficult to detect. The W boson is involved in the production and decay of neutrinos, which has led to a better understanding of their properties and interactions.

The W boson is an elementary particle that mediates the weak nuclear force, which is responsible for radioactive decay and nuclear fusion in stars. It is a gauge boson with a mass of approximately 80 GeV/c^2 and a spin of 1, and comes in two varieties, the positively charged W+ boson and the negatively charged W- boson. The W boson interacts with both leptons and quarks and can change the flavor of quarks by converting one type of quark into another type. The discovery of the W boson was a significant achievement in particle physics, as it provided experimental evidence for the electroweak theory and has played a crucial role in the study of neutrinos.

Chapter 5: Quantum Interpretations

Quantum mechanics is one of the most successful scientific theories ever developed, describing the behavior of subatomic particles with extraordinary accuracy. However, it is also one of the most perplexing and counterintuitive theories, leading to a number of different interpretations that attempt to explain its strange and sometimes contradictory predictions.

One of the most widely accepted interpretations of quantum mechanics is the Copenhagen interpretation, named after the city where it was first formulated. This interpretation holds that quantum systems do not have definite properties until they are measured, at which point the system "collapses" into a single, well-defined state. This interpretation is based on the idea that the act of measurement is fundamentally different from other physical processes, and that it somehow causes the quantum system to "choose" a particular state.

Another interpretation of quantum mechanics is the many-worlds interpretation, which was proposed by physicist Hugh Everett in the 1950s. According to this interpretation, every time a measurement is made, the universe splits into multiple "branches" in which the quantum system has different outcomes. In this view, the act of measurement does not collapse the system into a single state, but rather reveals the outcome of one of many possible quantum interactions.

A third interpretation of quantum mechanics is the pilot-wave theory, also known as the de Broglie-Bohm interpretation. This interpretation proposes that quantum particles are guided by a hidden "pilot wave" that determines their behavior in a deterministic manner. In this view, the wave function that describes the quantum system is not a complete description of reality, but rather a mathematical tool that describes the probability of different outcomes.

These are just a few examples of the many different interpretations of quantum mechanics that have been proposed over the years. Each interpretation attempts to reconcile the strange and counterintuitive predictions of quantum mechanics with our classical understanding of the world, but they often lead to radically different conclusions about the nature of reality.

One of the challenges of interpreting quantum mechanics is that many of its predictions are probabilistic in nature, meaning that they describe the likelihood of different outcomes rather than definite properties. This has led some physicists to question whether the theory can be considered a complete description of reality, or whether there are underlying physical processes that are not captured by the mathematics of quantum mechanics.

Despite the challenges and disagreements surrounding quantum interpretations, the theory remains one of the most powerful tools in modern science, with applications ranging from quantum computing to materials science to drug discovery. As researchers continue to explore the mysteries of quantum mechanics, we can expect to see new interpretations and insights emerge that will shed light on the fundamental nature of the universe.

The Copenhagen interpretation of quantum mechanics is one of the most widely accepted interpretations of the theory. It was developed by a group of physicists, including Niels Bohr and Werner Heisenberg, in the 1920s and 1930s in Copenhagen, Denmark. The interpretation holds that quantum systems do not have definite properties until they are measured, at which point the system "collapses" into a single, well-defined state. We will explore the key concepts and implications of the Copenhagen interpretation.

One of the fundamental tenets of the Copenhagen interpretation is the idea of complementarity, which holds that quantum systems can exhibit both wave-like and particle-like behavior, but not both at the same time. This means that the properties of a quantum system are not fixed until they are measured, at which point the system collapses into a single state that corresponds to one of the possible outcomes of the measurement.

The act of measurement is a crucial element of the Copenhagen interpretation, as it is the process that causes the system to collapse. According to the interpretation, the act of measurement is fundamentally different from other physical processes, such as the interactions between particles, and it somehow causes the quantum system to "choose" a particular state.

The Copenhagen interpretation also introduces the concept of the wave function, which is a mathematical function that describes the probability of finding a particle in a particular state. The wave function is used to calculate the probabilities of different outcomes of a measurement, and it is a central concept in quantum mechanics.

One of the key implications of the Copenhagen interpretation is that it introduces a fundamental element of uncertainty into quantum mechanics. Because the properties of a quantum system are not fixed until they are measured, it is impossible to predict the outcome of a measurement with complete certainty. Instead, quantum mechanics provides only probabilities for the different possible outcomes.

This uncertainty has led to a number of thought experiments and paradoxes, such as the famous Schrödinger's cat paradox, which illustrates the strange and counterintuitive predictions of quantum mechanics. In this paradox, a cat is placed in a sealed box with a radioactive source that may or may not decay, triggering a mechanism that kills the cat. According to quantum mechanics, the cat is neither alive nor dead until the box is opened and the state of the radioactive source is measured, at which point the cat's fate is determined.

The Copenhagen interpretation has been criticized by some physicists for its reliance on the concept of measurement and the lack of a clear explanation for how and why measurements cause the collapse of the quantum system. Additionally, the interpretation has been criticized for its apparent lack of realism, as it suggests that the properties of a quantum system are not fixed until they are measured.

Despite these criticisms, the Copenhagen interpretation remains one of the most widely accepted interpretations of quantum mechanics. It provides a powerful framework for understanding the probabilistic and counterintuitive predictions of quantum mechanics, and it has been the foundation of many of the practical applications of quantum mechanics, such as quantum computing and cryptography.

In conclusion, the Copenhagen interpretation of quantum mechanics is a powerful and widely accepted interpretation that has been instrumental in advancing our understanding of the quantum world. Although it has been subject to criticism and debate, it remains a foundational concept in modern physics, providing a framework for exploring the strange and counterintuitive predictions of quantum mechanics.

The Many-Worlds Interpretation (MWI) of quantum mechanics is a controversial interpretation that has garnered significant attention and debate in the field. The theory was first introduced by physicist Hugh Everett in 1957 in his doctoral thesis at Princeton University. The MWI suggests that the universe is constantly branching into multiple parallel universes, each representing a different outcome of a quantum measurement. We will explore the origins and implications of the Many-Worlds

Interpretation, as well as some of the criticisms and alternative interpretations of quantum mechanics.

The Many-Worlds Interpretation of quantum mechanics emerged from the work of Hugh Everett in the 1950s. Everett was attempting to solve what is known as the measurement problem in quantum mechanics. In classical mechanics, a measurement simply reveals the pre-existing value of a physical quantity, such as the position or momentum of a particle. However, in quantum mechanics, a measurement appears to collapse the wavefunction of a system, resulting in a single, definite outcome. This has led to much debate and controversy, with some interpretations suggesting that the act of measurement somehow creates the observed result.

Everett proposed a different solution to the measurement problem, suggesting that the wavefunction never collapses, but instead continues to evolve into multiple parallel universes. In other words, every time a quantum measurement is made, the universe splits into multiple parallel branches, each representing a different possible outcome of the measurement. This theory is based on the idea that quantum mechanics is a deterministic theory, meaning that the wavefunction evolves according to a set of mathematical rules that determine all possible outcomes of a measurement. The MWI posits that each of these outcomes actually occurs in a separate universe, so that there is no collapse of the wavefunction or destruction of information.

The Many-Worlds Interpretation has several implications for our understanding of quantum mechanics and the nature of reality itself. One of the most striking implications is that it leads to the conclusion that there are an infinite number of parallel universes, each representing a different possible outcome of a quantum measurement. This means that there are countless versions of ourselves, each living in a different universe that resulted from a different quantum measurement.

Another implication of the MWI is that it eliminates the need for wavefunction collapse, which has long been a puzzling aspect of quantum mechanics. In the Many-Worlds Interpretation, the wavefunction never collapses, but instead continues to evolve into multiple parallel universes, each representing a different possible outcome of a measurement. This means that the entire universe is described by a single wavefunction that includes all possible outcomes, rather than a collapsed wavefunction that represents only one outcome.

The Many-Worlds Interpretation of quantum mechanics has been met with significant criticism and debate since its inception. One of the main criticisms of the MWI is that it is difficult to test or confirm experimentally. Since the theory posits the existence of an infinite number of parallel universes, it is impossible to observe or interact with these other universes in any meaningful way.

Another criticism of the MWI is that it appears to violate Occam's razor, a principle of parsimony that suggests the simplest explanation is usually the correct one. The MWI posits the existence of an infinite number of parallel universes, which is a very complex and seemingly unnecessary explanation for the behavior of quantum systems.

The pilot wave theory, also known as the de Broglie-Bohm theory or Bohmian mechanics, is an interpretation of quantum mechanics that suggests that particles are guided by a "pilot wave" that determines their behavior. This theory was developed by Louis de Broglie and David Bohm in the 1920s and 1950s, respectively, and is considered a deterministic interpretation of quantum mechanics.

The pilot wave theory is built on the foundations of wave-particle duality, which is a fundamental principle of quantum mechanics. The theory proposes that all particles, including electrons and photons, have both wave-like and particle-like properties, and their behavior can be described using probability distributions.

According to the pilot wave theory, every quantum system is guided by a pilot wave, which is a wave function that determines the particle's behavior. The wave function is described by the Schrödinger equation, which is a fundamental equation in quantum mechanics that describes the evolution of the wave function over time.

The Schrödinger equation can be written as:

$$i\hbar \partial \Psi / \partial t = H\Psi$$

where $\hbar$ is the reduced Planck constant, Ψ is the wave function, t is time, and H is the Hamiltonian operator. The Hamiltonian operator represents the total energy of the system and includes information about the particles' positions and momenta.

In the pilot wave theory, the wave function is not just a mathematical construct but has a physical reality. The wave function describes the motion of the pilot wave, which guides the particles' behavior. The particles themselves are considered to have a definite position and momentum, but their behavior is determined by the pilot wave

The pilot wave theory offers a deterministic interpretation of quantum mechanics, which means that it suggests that particles have a definite position and momentum at all times. This is in contrast to the Copenhagen interpretation, which proposes that particles do not have a definite position or momentum until they are measured.

One of the key concepts of the pilot wave theory is non-locality. Non-locality refers to the idea that particles can influence each other instantaneously, regardless of their distance apart. This is because the pilot wave is considered to be a single entity that guides the behavior of all particles in the system.

The pilot wave theory also suggests that particles have a hidden variable, which determines their behavior. This hidden variable is not observable and cannot be measured directly but is considered to be the guiding force behind the particles' behavior.

The pilot wave theory has several implications for quantum mechanics and our understanding of the physical world. It suggests that there is a physical reality underlying the wave function, and particles have a definite position and momentum at all times. It also challenges the idea of the observer effect, which suggests that particles behave differently when they are observed.

The pilot wave theory has received criticism from some physicists, who argue that it does not fully explain certain quantum phenomena, such as entanglement and the double-slit experiment. These phenomena are difficult to explain using a deterministic interpretation of quantum mechanics and require the use of probability distributions.

The pilot wave theory also requires the existence of a preferred frame of reference, which is not compatible with the principles of relativity. This means that the theory cannot be applied to all physical systems and is limited in its scope.

Chapter 6: Quantum Field Theory

Quantum Field Theory (QFT) is a fundamental theory in physics that describes the behavior of elementary particles and their interactions. It is based on the concept of quantum fields, which are fields that exist everywhere in space and time and describe the behavior of elementary particles.

In classical physics, fields are continuous, smooth functions that describe the behavior of physical quantities, such as the electric and magnetic fields. In QFT, however, fields are quantized, which means that they are composed of discrete particles called quanta. These quanta are the elementary particles, such as electrons and quarks, that make up matter.

QFT describes the behavior of particles and their interactions using a set of mathematical equations called the Lagrangian. The Lagrangian describes the energy and momentum of the particles and the forces that govern their interactions. The Lagrangian is a function of the fields and their derivatives, and it describes how the fields interact with each other and with the particles that they describe.

One of the key concepts in QFT is the concept of the vacuum state. In classical physics, the vacuum is defined as the absence of any matter or energy. In QFT, however, the vacuum is not empty but is filled with a sea of virtual particles. These virtual particles are constantly being created and destroyed by the fields, and they can interact with real particles to produce observable effects.

Another important concept in QFT is the renormalization process. When calculating the behavior of particles and their interactions using QFT, infinite quantities can arise in the calculations. These infinities need to be removed by a process called renormalization, which involves adjusting the parameters in the Lagrangian to account for the infinities.

QFT is a very powerful theory that has been very successful in explaining many fundamental physical phenomena, such as the behavior of subatomic particles and the structure of matter. It is also used in many other areas of physics, such as particle physics, condensed matter physics, and cosmology.

In summary, QFT is a fundamental theory in physics that describes the behavior of particles and their interactions using the concept of quantum fields. The theory is based on the Lagrangian, which describes the energy and momentum of the particles and the forces that govern their interactions. The theory is very successful in explaining many fundamental physical phenomena and is used in many areas of physics.

Electromagnetic fields are a fundamental aspect of the universe, describing the way that electric and magnetic fields interact with one another to create a wide range of physical phenomena. This paper will explore the basic concepts behind electromagnetic fields, their mathematical description, and some of the practical applications of this field of study.

The theory of electromagnetism is based on the idea that electric charges create electric fields, and moving electric charges create magnetic fields. Electromagnetic fields are the combination of these two types of fields. Electromagnetic waves are created when electric and magnetic fields oscillate together, creating a wave-like pattern that can travel through space.

Mathematically, electromagnetic fields are described using Maxwell's equations. These equations describe the behavior of electric and magnetic fields, as well as how they interact with one another. There are four Maxwell's equations that describe the behavior of electromagnetic fields:

Gauss's law for electric fields: This equation states that the electric flux through a closed surface is proportional to the electric charge enclosed by the surface.

Gauss's law for magnetic fields: This equation states that there are no magnetic charges, and that the magnetic flux through a closed surface is always zero.

Faraday's law of electromagnetic induction: This equation states that a changing magnetic field induces an electric field.

Ampere's law with Maxwell's correction: This equation states that a changing electric field induces a magnetic field, and that magnetic fields are produced by electric currents.

Together, these equations describe the behavior of electromagnetic fields and how they propagate through space.

Electromagnetic fields have many practical applications, from electricity and magnetism in everyday life to more advanced technologies. For example, electric motors use electromagnetic fields to convert electrical energy into mechanical energy, and generators use the opposite process to convert mechanical energy into electrical energy. Electromagnetic waves, such as radio waves, microwaves, and visible light, are used for communication, sensing, and many other applications.

In addition to their practical applications, electromagnetic fields also play a critical role in modern physics. For example, quantum field theory describes the behavior of particles and their interactions using electromagnetic fields as one of its fundamental building blocks. Similarly, the study of particle physics relies heavily on the behavior of electromagnetic fields, as particles such as electrons and protons are charged and interact with electromagnetic fields.

In conclusion, electromagnetic fields are a fundamental aspect of the universe, describing the way that electric and magnetic fields interact with one another to create a wide range of physical phenomena. The behavior of these fields is described mathematically using Maxwell's equations, and they have many practical applications, as well as being critical to the understanding of modern physics.

Gravitational fields are a fundamental aspect of the universe, describing the way that massive objects interact with one another to create the force of gravity.

The theory of gravity is based on the idea that objects with mass create a gravitational field that exerts a force on other objects with mass. This force is proportional to the masses of the objects and inversely proportional to the distance between them. The mathematical description of gravitational fields is provided by

Einstein's theory of general relativity, which describes gravity as the curvature of spacetime caused by the presence of mass and energy.

The strength of a gravitational field is determined by the mass of the object creating the field. The larger the mass of the object, the stronger the gravitational field it creates. This is why the gravitational field of the Earth is much stronger than that of the Moon, despite the fact that the Moon is much closer to the Earth than any other celestial body.

Gravitational fields have many practical applications in the field of astronomy. For example, astronomers can use gravitational lensing to observe distant objects that would otherwise be obscured by other objects in their line of sight. Gravitational lensing occurs when the gravitational field of a massive object, such as a galaxy or a black hole, bends the path of light passing through it. This can cause light from a distant object to be bent around the massive object, creating a magnified and distorted image of the object.

In addition to their practical applications, gravitational fields also play a critical role in modern physics. General relativity, which describes the behavior of gravitational fields, is one of the two pillars of modern physics, along with quantum mechanics. The study of black holes, which are objects with such strong gravitational fields that not even light can escape them, is also a critical area of research in modern physics.

General relativity is a theory of gravitation that was developed by Albert Einstein in 1915. It is a foundational theory of modern physics and describes the behavior of massive objects in the presence of gravitational fields. This paper will provide an overview of the key concepts behind general relativity, its mathematical description, and some of its practical applications.

The central concept of general relativity is that gravity is not a force that objects exert on each other, but rather a curvature of spacetime caused by the presence of mass and energy. In other words, objects with mass create a gravitational field that causes the curvature of spacetime in their vicinity. The curvature of spacetime, in turn, affects the motion of other objects in the vicinity, including light.

The mathematical description of general relativity involves the use of tensors and differential equations. The theory is based on the idea that the curvature of spacetime is determined by the distribution of mass and energy in the universe. The mathematical equations of general relativity are complex and non-linear, making it difficult to solve them exactly for all but the simplest situations. Nevertheless, the theory has been tested and confirmed by numerous experiments and observations, including the bending of light around massive objects and the precession of the orbit of Mercury.

One of the most important predictions of general relativity is the existence of black holes. Black holes are objects with such strong gravitational fields that not even light can escape them. They are formed when a massive star collapses under the force of its own gravity, creating a singularity – a point of infinite density and zero volume. Black holes are invisible, but their presence can be inferred from the effects of their gravity on nearby matter and light.

Nicholas Tassy-Becz

General relativity has many practical applications, including the prediction and detection of gravitational waves. Gravitational waves are ripples in spacetime caused by the acceleration of massive objects, such as colliding black holes or neutron stars. They were first predicted by general relativity in 1916, but it was not until 2015 that they were directly detected by the Laser Interferometer Gravitational-Wave Observatory (LIGO).

Another important application of general relativity is in the field of cosmology. The theory predicts that the universe is expanding and that the expansion is accelerating. This prediction has been confirmed by observations of distant galaxies and the cosmic microwave background radiation.

The electromagnetic field theory describes the behavior of photons, which are the fundamental particles of light and other electromagnetic radiation. The theory also describes the behavior of charged particles, such as electrons, that interact with the electromagnetic field.

The electromagnetic field theory is based on the principle of gauge invariance, which states that the equations of motion for the electromagnetic field must be invariant under certain transformations. These transformations correspond to changes in the phase of the wave function describing the electromagnetic field. The equations of motion for the electromagnetic field are derived from a Lagrangian density, which specifies the energy and interactions of the electromagnetic field.

Gravitational fields, on the other hand, arise from the curvature of spacetime in the presence of massive objects. General relativity, which is a theory of gravity, is also a subset of quantum field theory. In general relativity, the curvature of spacetime is described by the metric tensor, which is a mathematical object that describes the geometry of spacetime. The metric tensor is derived from a Lagrangian density that specifies the energy and interactions of the gravitational field.

In quantum field theory, gravitational fields are described by the theory of quantum gravity. However, quantum gravity is currently an incomplete theory and remains an active area of research. Despite this, some progress has been made in developing theories of quantum gravity, such as string theory and loop quantum gravity.

One of the challenges of developing a theory of quantum gravity is that it requires reconciling the principles of quantum mechanics and general relativity, which are both highly successful theories in their respective domains. In particular, the principles of general relativity imply that spacetime is continuous and deterministic, while quantum mechanics implies that particles can exist in multiple states simultaneously and are inherently unpredictable.

One proposed solution to this challenge is the concept of a quantum field theory of gravity. This approach treats the gravitational field as a quantum field, much like the electromagnetic field. In this approach, particles that carry gravitational force, called gravitons, are the fundamental excitations of the quantum gravitational field.

In summary, quantum field theory provides a framework for describing the behavior of subatomic particles in the presence of fields. The electromagnetic field theory is a subset of quantum field theory that describes the behavior of photons and charged particles in electromagnetic fields. General relativity is also a subset of quantum field theory that describes the curvature of spacetime in the presence of massive objects. Quantum gravity is a proposed theory of quantum field theory that aims to reconcile the principles of quantum mechanics and general relativity by treating the gravitational field as a quantum field.

The quantum field theory standard model is a theoretical framework that describes the behavior of elementary particles and their interactions. It is based on the principles of quantum mechanics and special relativity and is the most widely accepted theory of particle physics to date. In this paper, we will discuss the quantum field theory standard model in detail.

The quantum field theory standard model is a combination of three theories: the electromagnetic theory, the weak interaction theory, and the strong interaction theory. These three theories are unified through the use of gauge symmetry, which is a mathematical principle that ensures the consistency of the theory.

The electromagnetic theory describes the behavior of electrically charged particles and their interactions with the electromagnetic field. It is based on the principle of gauge invariance and is described by the quantum electrodynamics (QED) theory.

The weak interaction theory describes the behavior of particles that are involved in weak interactions, such as beta decay. Weak interactions are mediated by the exchange of particles called W and Z bosons. The weak interaction theory is described by the electroweak theory, which is a combination of the electromagnetic theory and the weak interaction theory.

The strong interaction theory describes the behavior of particles that are involved in strong interactions, such as quarks and gluons. Strong interactions are mediated by the exchange of particles called gluons. The strong interaction theory is described by quantum chromodynamics (QCD).

The quantum field theory standard model also includes the Higgs boson, which is a particle that is responsible for giving other particles mass. The Higgs boson is described by the Higgs mechanism, which involves the spontaneous breaking of gauge symmetry.

In the quantum field theory standard model, particles are described as excitations of quantum fields. Each particle corresponds to a particular excitation of a particular quantum field. For example, the photon is the excitation of the electromagnetic field, while the W and Z bosons are excitations of the electroweak field.

Particles are classified into two broad categories: fermions and bosons. Fermions are particles that have half-integer values of spin, such as electrons, quarks, and neutrinos. Bosons are particles that have integer values of spin, such as photons, W and Z bosons, and the Higgs boson.

The behavior of particles in the quantum field theory standard model is described by the Feynman diagrams. Feynman diagrams are graphical representations of particle interactions and are used to calculate the probability amplitudes for a given particle interaction.

The quantum field theory standard model has been extensively tested and is one of the most successful theories in physics. It has accurately predicted the behavior of particles and their interactions in a wide range of experiments. However, the quantum field theory standard model is not complete and has some limitations. For example, it does not incorporate gravity and

At the heart of the QFTSM is the quantum field, which describes the behavior of a particle at all points in space and time. The QFTSM uses a Lagrangian density to describe the dynamics of the quantum field. The Lagrangian density is a function that describes the energy of a system in terms of its position and velocity.

The QFTSM equation is a set of mathematical equations that describe the behavior of the quantum field. The QFTSM equation is based on the principles of gauge symmetry, which is a mathematical principle that ensures the consistency of the theory.

The QFTSM equation has several components, including the Dirac equation, the Maxwell equations, and the Yang-Mills equations. The Dirac equation describes the behavior of fermions, which are particles that have half-integer values of spin, such as electrons, quarks, and neutrinos. The Maxwell equations describe the behavior of the electromagnetic field, while the Yang-Mills equations describe the behavior of the strong and weak nuclear forces.

The QFTSM equation also includes the Higgs field, which is a scalar field that is responsible for giving other particles mass. The Higgs field is described by the Higgs mechanism, which involves the spontaneous breaking of gauge symmetry.

The QFTSM equation uses the concept of a gauge boson to describe the force carriers that mediate the interactions between particles. Gauge bosons are particles that have integer values of spin, such as photons, W and Z bosons, and gluons. The interactions between particles are mediated by the exchange of gauge bosons.

The QFTSM equation uses a technique called renormalization to account for the effects of virtual particles, which are particles that are not directly observed but affect the behavior of other particles. Renormalization involves the use of mathematical techniques to cancel out the effects of virtual particles and ensure that the QFTSM equation remains consistent.

The QFTSM equation is a complex mathematical framework that has been extensively tested and is one of the most successful theories in physics. It has accurately predicted the behavior of particles and their interactions in a wide range of experiments. However, the QFTSM equation is not complete and has some limitations. For example, it does not incorporate gravity and is unable to describe the behavior of particles at extremely high energies.

In conclusion, the QFTSM equation is a mathematical framework that describes the behavior of particles and their interactions. The QFTSM equation is based on the principles of quantum mechanics and special relativity and is the most widely accepted theory of particle physics to date. The QFTSM equation uses the concept of gauge symmetry and the exchange of gauge bosons to describe the interactions between particles. The QFTSM equation is a complex mathematical framework that has been extensively tested and is one of the most successful theories in physics

Chapter 7: Quantum computing

Quantum computing is an emerging field of computer science that aims to harness the principles of quantum mechanics to perform complex calculations more efficiently than classical computers. In this paper, we will discuss the fundamentals of quantum computing, including its underlying principles, hardware, and potential applications.

At the heart of quantum computing is the concept of a qubit, which is the basic unit of quantum information. Unlike classical bits, which can only be in a state of 0 or 1, qubits can exist in a superposition of both states simultaneously. This property allows quantum computers to perform many calculations simultaneously, which is known as quantum parallelism.

Another key concept in quantum computing is entanglement, which occurs when two qubits become correlated in a way that cannot be described by classical physics. Entanglement allows quantum computers to perform certain calculations more efficiently than classical computers.

To manipulate and measure qubits, quantum computers use a variety of technologies, including superconducting circuits, trapped ions, and topological qubits. These technologies are designed to isolate the qubits from the environment and prevent them from being disturbed by external factors, which can cause errors in calculations.

One of the most well-known quantum algorithms is Shor's algorithm, which can be used to factor large numbers more efficiently than classical algorithms. This is a significant breakthrough because many encryption techniques used today rely on the difficulty of factoring large numbers, and quantum computers could potentially break these encryption methods.

Other potential applications of quantum computing include simulations of complex physical systems, optimization problems, and machine learning. For example, quantum computers could be used to simulate the behavior of molecules, which could lead to the development of new materials and drugs.

Despite the potential benefits of quantum computing, there are several challenges that must be overcome before it can become a practical technology. One of the biggest challenges is the problem of quantum decoherence, which occurs when the qubits become entangled with their environment, causing them to lose their quantum properties. Researchers are working to develop error correction techniques to mitigate the effects of decoherence.

In conclusion, quantum computing is an emerging field of computer science that has the potential to revolutionize computing by exploiting the principles of quantum mechanics. Quantum computers use qubits, which can exist in a superposition of states, and entanglement, which allows for efficient parallelism in calculations. While there are still significant challenges to be addressed, quantum computing has the potential to solve complex problems that are currently intractable using classical computers, and could lead to the development of new materials, drugs, and encryption techniques.

Quantum computers have the potential to revolutionize computing by providing exponential speedups for certain types of calculations. However, building a functional quantum computer is a complex and challenging task that requires the use of specialized materials and conditions. In this paper, we will discuss the materials and conditions needed to build a quantum computer.

The heart of a quantum computer is the qubit, which is the basic unit of quantum information. Qubits can be implemented using a variety of physical systems, including superconducting circuits, trapped ions, and topological qubits. Each of these systems requires different materials and conditions to function properly.

Superconducting circuits are a popular choice for implementing qubits in quantum computers. These circuits consist of thin films of superconducting material, such as niobium, that are patterned into the shape of a qubit. Superconducting qubits require extremely low temperatures, typically around 20 millikelvin (-273 degrees Celsius), to function properly. This requires the use of specialized cryogenic equipment, such as dilution refrigerators, which can maintain these low temperatures for extended periods of time. Trapped ions are another popular choice for implementing qubits in quantum computers. In this system, ions are trapped using electromagnetic fields and manipulated using laser pulses. Trapped ion qubits require very high vacuum conditions to prevent the ions from interacting with air molecules, which can cause decoherence and errors in calculations. Additionally, the lasers used to manipulate the ions must be carefully controlled to prevent unwanted interactions between the ions and the environment. Topological qubits are a newer approach to quantum computing that uses the properties of exotic materials to create qubits. These materials include topological insulators, which have unique properties that make them resistant to decoherence. Topological qubits can operate at higher temperatures than other qubit technologies, but they require specialized fabrication techniques to create the necessary structures.

In addition to specialized materials, building a quantum computer also requires precise control over the qubits and their environment. This is typically accomplished using feedback control systems that measure and adjust the qubits to maintain their quantum states. These control systems require specialized electronics and software to operate effectively.

Another important consideration when building a quantum computer is the connectivity between qubits. Quantum computers require a large number of qubits to perform useful calculations, and these qubits must be connected in a way that allows them to communicate with each other. This requires the use of specialized wiring and control systems that can route signals between qubits without causing interference or errors.

In conclusion, building a quantum computer requires the use of specialized materials and conditions, including cryogenic equipment, vacuum chambers, lasers, and exotic materials. Precise control over the qubits and their environment is also essential, and this is typically accomplished using feedback control systems and specialized electronics and software. Despite the challenges involved in building a quantum computer, advances in qubit technologies and control systems are bringing us closer to realizing the potential of quantum computing.

Quantum gates are the building blocks of quantum circuits, which are the equivalent of classical circuits in quantum computing. Quantum gates operate on qubits to manipulate their quantum states, and they are described by unitary matrices in quantum mechanics. These matrices have the property that their inverse is equal to their conjugate transpose, which ensures that quantum gates preserve the normalization of the wave function.

One of the most famous quantum gates is the Hadamard gate, which is used to create superpositions of qubits. The Hadamard gate is described by the following matrix:

Copy code

```
1 1
1 -1
```

When applied to a qubit in the 0 state, the Hadamard gate creates a superposition of 0 and 1:

scss
Copy code

```
|0) -> (|0) + |1)) / sqrt(2)
```

Another important quantum gate is the CNOT gate, which is used to create entangled states of qubits. The CNOT gate is described by the following matrix:

Copy code

1 0 0 0
0 1 0 0
0 0 0 1
0 0 1 0

The CNOT gate operates on two qubits, with the first qubit acting as the control qubit and the second qubit acting as the target qubit. When the control qubit is in the state 1, the CNOT gate flips the state of the target qubit. This creates an entangled state between the two qubits, which is a superposition of both qubits being in the state 0 or both qubits being in the state 1.

Quantum algorithms are algorithms that are designed to run on a quantum computer. These algorithms take advantage of the unique properties of qubits to solve certain types of problems much faster than classical algorithms. One of the most famous quantum algorithms is Shor's algorithm, which is used to factor large numbers into their prime factors. Shor's algorithm provides an exponential speedup over classical algorithms, which has important implications for cryptography and other fields.

In conclusion, the mathematics behind quantum computers is based on the principles of quantum mechanics, which describe the behavior of particles at the atomic and subatomic level. Quantum gates are used to manipulate the quantum states of qubits, and quantum algorithms are designed to take advantage of the unique properties of qubits to solve certain types of problems much faster than classical algorithms. Despite the challenges involved in building and operating quantum computers, advances in qubit technologies and quantum algorithms are bringing us closer to realizing the potential of quantum computing. The development of the first quantum computers is a story of scientific innovation and technical ingenuity. In this paper, we will discuss the history of the first quantum computers, including the theoretical and experimental breakthroughs that made them possible. The idea of a quantum computer was first proposed by the physicist Richard Feynman in 1982. Feynman realized that a classical computer would have difficulty simulating quantum systems, which motivated him to propose the idea of a quantum computer that could perform such simulations efficiently. However, it was not until the 1990s that the first experimental proposals for quantum computers were put forward.

The first experimental realization of a quantum computer was achieved by a team of researchers at IBM in 1998. The team built a quantum computer using nuclear magnetic resonance (NMR) techniques, which allowed them to manipulate the spin states of atomic nuclei in a molecule. The quantum computer consisted of two qubits, and the team used it to perform simple quantum algorithms, such as the Deutsch-Jozsa algorithm and the Grover search algorithm.

The NMR quantum computer was limited in its capabilities, however, due to the difficulty of scaling up the number of qubits. A major breakthrough in quantum computing came in 2000, when a team of researchers at the University of California, Santa Barbara, led by John Martinis, developed a superconducting qubit that was much more robust than previous qubits. The superconducting qubit was made from a tiny loop of superconducting wire, and it was able to maintain its quantum state for much longer than previous qubits.

The development of superconducting qubits led to the creation of the first scalable quantum computing systems. In 2007, a team of researchers at the University of Waterloo built a four-qubit superconducting quantum computer, which they used to run the Shor's algorithm for factorizing the number 15. The researchers were able to factorize the number using the quantum computer, which was a major milestone in the field of quantum computing.

Since then, the field of quantum computing has seen rapid progress, with companies such as IBM, Google, and Microsoft investing heavily in the development of quantum computing technologies. In 2019, Google announced that it had achieved quantum supremacy, which means that its quantum computer had performed a calculation that would be infeasible for a classical computer. The calculation involved generating random numbers, which is a problem that is well-suited to the capabilities of a quantum computer.

In conclusion, the development of the first quantum computers has been a remarkable achievement in the history of science and technology. The theoretical and experimental breakthroughs that made them possible have opened up new possibilities for computing and for the study of quantum systems. While quantum computers are still in the early stages of development, they have already shown the potential to revolutionize fields such as cryptography, optimization, and machine learning. As researchers continue to develop new qubit technologies and quantum algorithms, we can expect to see even more exciting developments in the field of quantum computing in the years to come.

Drug discovery is a complex process that involves the identification and optimization of chemical compounds that can effectively target specific diseases. Traditionally, this process has been time-consuming and expensive, as it requires screening large numbers of compounds and optimizing their properties using trial and error methods. However, recent developments in quantum computing have opened up new possibilities for accelerating the drug discovery process.

One of the key advantages of quantum computing for drug discovery is its ability to simulate complex molecular systems. The behavior of molecules is governed by the laws of quantum mechanics, which can be difficult to model using classical computers. Quantum computers, on the other hand, are designed to work with quantum systems and can perform simulations that are not feasible on classical computers.

Quantum computers can be used to simulate the behavior of molecules by solving the Schrödinger equation, which describes the quantum states of the system. This allows researchers to predict the properties of chemical compounds, such as their stability, reactivity, and binding affinity to specific biological targets. By simulating the behavior of large numbers of compounds, quantum computers can help to identify promising drug candidates that can be further optimized in the laboratory.

In addition to molecular simulations, quantum computers can also be used to optimize the properties of chemical compounds. Optimization involves adjusting the structure of a compound to improve its properties, such as its binding affinity or solubility. This can be done using machine learning algorithms that are trained on large datasets of chemical compounds. By taking into account the quantum properties of the molecules, such as their electronic structure and quantum interactions, these algorithms can make more accurate predictions about the properties of new compounds.

There are several challenges that need to be overcome in order to use quantum computers for drug discovery. One of the biggest challenges is the need for error correction, as quantum systems are highly susceptible to noise and decoherence. Current quantum computers have a limited number of qubits and a high error rate, which makes it difficult to perform accurate simulations of large molecules. However, researchers are working on developing new error correction techniques and more powerful quantum computers, which could overcome these challenges in the future.

Despite these challenges, there have been some promising results in the field of quantum computing for drug discovery. In 2020, researchers at the University of Toronto used a quantum computer to predict the binding affinity of a protein to a small molecule drug. The prediction was verified in the laboratory, demonstrating the potential of quantum computing for drug discovery.

In conclusion, quantum computing has the potential to revolutionize the field of drug discovery by enabling faster and more accurate simulations of molecular systems. While there are still several challenges that need to be addressed, the development of new qubit technologies and error correction techniques could enable the creation of more powerful quantum computers that can be used for drug discovery applications. As researchers continue to explore the capabilities of quantum computers, we can expect to see even more exciting developments in the field of drug discovery in the years to com

Chapter 8: The Early Days of the Universe

The Big Bang Theory is a scientific model that describes the origin and evolution of the universe. The theory suggests that the universe began as a single point or singularity, and then rapidly expanded and cooled down, leading to the formation of matter and energy as we know it today.

The Big Bang Theory was first proposed in the early 20th century by Belgian astronomer Georges Lemaître, who suggested that the universe had a beginning and was expanding. His theory was supported by observations made by astronomer Edwin Hubble, who found that galaxies were moving away from each other, suggesting that the universe was expanding.

The most widely accepted model of the Big Bang suggests that the universe began as a singularity, a point of infinite density and temperature. At this point, the laws of physics as we know them to break down, making it difficult to describe the conditions that existed during the earliest moments of the universe. However, as the universe rapidly expanded and cooled down, it entered a period known as inflation, during which it expanded faster than the speed of light. This expansion led to the formation of subatomic particles such as protons, neutrons, and electrons, as well as photons, the particles that makeup light. As the universe continued to cool down, these particles began to combine and form atoms, which eventually led to the formation of stars and galaxies. The early universe was primarily made up of hydrogen and helium, and it wasn't until stars began to form and die that heavier elements such as carbon and oxygen were created. One of the key pieces of evidence for the Big Bang Theory is the cosmic microwave background radiation, which is thought to be leftover radiation from the early universe. This radiation was first detected in the 1960s by astronomers Arno Penzias and Robert Wilson, and it has been studied extensively by scientists in the decades since. While the Big Bang Theory is widely accepted among scientists, there are still many questions about the earliest moments of the universe that remain unanswered. Some scientists are working on developing new models that could provide more insight into the conditions that existed during the Big Bang, while others are focused on using observations of the universe to better understand its evolution over time In physics, a singularity refers to a point in space-time where physical quantities become infinite or indeterminate. Singularities are points in space-time where the laws of physics break down, and our current understanding of the universe is unable to provide a satisfactory explanation for their existence. Singularities are believed to exist at the center of black holes and at the beginning of the universe during the Big Bang.

The most famous singularity is the one that exists at the center of a black hole. According to general relativity, the mass of a black hole is compressed into a single point, known as a singularity, where the gravitational pull becomes so strong that not even light can escape its grasp. In this case, the laws of physics as we know them break down, and we need a theory of quantum gravity to understand the behavior of matter and energy at such a small scale.

The other famous singularity is the one that existed at the beginning of the universe, during the Big Bang. According to the prevailing cosmological model, the universe began as a singularity, a point of infinite density and temperature. At this point, the laws of physics as we know them break down, and we need a theory of quantum gravity to understand the behavior of matter and energy at such a small scale. The existence of singularities presents a major problem for physicists, as our current theories are unable to explain them. In general relativity, singularities are seen as a breakdown of the theory and indicate that new physics is needed. The search for a theory of quantum gravity is an active area of research, as it is hoped that such a theory will be able to explain the behavior of matter and energy at the smallest scales, including the behavior of singularities.

One proposed solution to the singularity problem is the idea of a "cosmic censorship hypothesis," which suggests that singularities are always hidden from view by an "event horizon" – a boundary beyond which nothing can escape the gravitational pull of a black hole. According to this hypothesis, singularities are always hidden from view, and we cannot observe them directly.

In conclusion, singularities represent a fundamental problem in physics, as our current theories are unable to explain their existence. The search for a theory of quantum gravity is ongoing, and it is hoped that such a theory will be able to explain the behavior of matter and energy at the smallest scales, including the behavior of singularities.

One of the most popular theories to explain what came before the Big Bang is the cyclic model. This theory suggests that the universe undergoes an infinite series of expansions and contractions, with each cycle beginning with a Big Bang and ending with a Big Crunch. According to this theory, the Big Bang was not the beginning of everything but was just the start of the current cycle of the universe.

Another theory proposed to explain what happened before the Big Bang is the concept of a multiverse. This theory suggests that there are multiple universes, each with its own set of physical laws and constants. In this view, the Big Bang was just the start of our own universe, while other universes may have had different beginnings or no beginning at all.

A third theory suggests that time itself did not exist before the Big Bang. This view is based on the idea that time is a product of the universe, and without the universe, time simply does not exist. In this theory, the question of what came before the Big Bang is meaningless since there was no "before" in the absence of time.

Regardless of the theories proposed, it is still unclear what exactly happened before the Big Bang, and further research is necessary to understand the origins and evolution of the universe. The study of the cosmic microwave background radiation, gravitational waves, and other cosmological observations provide clues to understanding the universe's early stages, and new advances in technology and scientific techniques will continue to expand our knowledge of the universe's origins.

Gravitational waves, ripples in spacetime caused by accelerating massive objects, offer new insights into the universe's earliest moments and how it evolved to its present state.

The universe was initially so hot and dense that it was not possible for matter to exist in its usual form. Instead, it was a plasma of high-energy particles and radiation. However, about 380,000 years after the Big Bang, the universe cooled enough for protons and electrons to combine and form neutral hydrogen atoms. This event is called the recombination era, and it is the point at which the universe became transparent to light.

The universe's early moments are still a mystery, and conventional electromagnetic waves cannot penetrate the plasma and provide information about it. However, gravitational waves are able to penetrate the plasma, providing a new window into the early universe's physics.

Gravitational waves are produced by the acceleration of massive objects, such as black holes or neutron stars. In the early universe, there were many massive objects accelerating rapidly, such as the hot plasma itself. Therefore, gravitational waves would have been produced during this time.

Cosmic microwave background radiation (CMB) is the leftover heat from the Big Bang that permeates the universe. It is essentially the radiation that was released when the universe became transparent to light during the recombination era. The CMB is a crucial piece of evidence supporting the Big Bang theory, and it has been studied extensively.

Gravitational waves can interact with the CMB, causing a phenomenon called gravitational lensing. Gravitational lensing is the bending of light due to the presence of a massive object. In the case of the CMB, gravitational lensing can be caused by the gravitational waves produced during the universe's early moments.

The study of gravitational waves provides new insights into the early universe's physics and the origins of the universe. Gravitational waves can penetrate the plasma that existed in the universe's early moments, providing information that conventional electromagnetic waves cannot. By studying the interaction between gravitational waves and the CMB, we can learn more about the universe's earliest moments and how it evolved to its present state. The Big Bang theory and the theory of gravitational waves are complementary, and studying them together can help us understand the universe's origins and evolution better. Gravitational waves were predicted by Albert Einstein in his theory of general relativity, which describes the behavior of gravity as a curvature of spacetime. According to this theory, massive objects create ripples in the fabric of spacetime as they move, similar to how a rock thrown into a pond creates waves on the surface. These ripples are called gravitational waves and they carry information about the motion of massive objects, such as colliding black holes or neutron stars. Gravitational waves are extremely difficult to detect because they are very weak and interact very weakly with matter. The first successful detection of gravitational waves was announced by the Laser Interferometer Gravitational-wave Observatory (LIGO) in 2016, which used two detectors located in the United States to observe the merger of two black holes over a billion light years away.

The basic principle behind LIGO and other gravitational wave detectors is interferometry, which involves splitting a laser beam into two perpendicular arms, bouncing the beams off mirrors, and then recombining them to detect any interference patterns caused by changes in the lengths of the arms. A passing gravitational wave causes the arms to lengthen and contract slightly, which changes the interference pattern and allows the detection of the wave. LIGO and other gravitational wave detectors are designed to be very sensitive to detect even the slightest changes in the length of the arms. They use a variety of technologies, such as high-power lasers, suspended mirrors, and advanced vibration isolation systems, to minimize noise and improve the signal-to-noise ratio. Since the first detection of gravitational waves in 2016, several other detections have been made by LIGO and other detectors around the world, including the Virgo detector in Italy and the KAGRA detector in Japan. These detections have provided new insights into the properties of black holes, neutron stars, and the universe as a whole, and have opened up a new window into studying the universe.

In conclusion, the detection of gravitational waves is a remarkable achievement in physics and astronomy. The ability to observe these waves opens up a new way to study the universe and provides a unique opportunity to test fundamental physics theories.

The Quantum Frontier

With the development of new and improved detectors, we can expect even more exciting discoveries in the years to come.

Chapter 9: Elements of the Periodic Table

In the last chapters of this book we have mostly covered the extremely small elements but now I will cover the forces driving atoms and thier interactions with subatomic particles.

Hydrogen, the simplest and most abundant element in the universe, has played a pivotal role in the development of quantum mechanics. The study of hydrogen and its properties has been instrumental in advancing our understanding of quantum physics and has led to some of the most fundamental discoveries in this field.In the early 20th century, scientists were beginning to explore the behavior of atoms, particularly the way in which they emitted and absorbed light. One of the most striking features of atomic spectra was the presence of discrete lines rather than a continuous spectrum. It was soon realized that these lines were the result of electrons in atoms jumping between discrete energy levels. Niels Bohr was the first to propose a model for the hydrogen atom that explained the discrete spectral lines. According to his model, the electron in a hydrogen atom could only occupy certain discrete energy levels, and when it jumped between these levels, it emitted or absorbed energy in the form of a photon. This was a major breakthrough in the field of quantum mechanics and laid the groundwork for the development of the Schrödinger equation.

The Schrödinger equation describes the behavior of quantum systems, including atoms, and has been instrumental in predicting the properties of hydrogen. The equation describes the probability density of finding an electron at a particular position around the nucleus of the atom, and it has been used to calculate the energy levels of the hydrogen atom and the wavelengths of the spectral lines it emits. The Schrödinger equation predicts that the energy levels of the hydrogen atom are quantized, meaning they can only take on certain discrete values. These energy levels are determined by the principal quantum number, which can take on integer values from 1 to infinity. The energy of the hydrogen atom is lowest when the electron is in the ground state, corresponding to the lowest energy level, and increases as the electron occupies higher energy levels. One of the most important applications of quantum mechanics to hydrogen is in the calculation of the hydrogen molecule. The hydrogen molecule is made up of two hydrogen atoms, and the Schrödinger equation can be used to calculate the energy levels and wave functions of the molecule. This has been instrumental in predicting the properties of molecules and has led to a greater understanding of chemical bonding. In conclusion, hydrogen has played a pivotal role in the development of quantum mechanics. The study of hydrogen and its properties has led to some of the most fundamental discoveries in this field and has laid the groundwork for our understanding of quantum mechanics. The Schrödinger equation has been instrumental in predicting the properties of hydrogen and has led to a greater understanding of the behavior of atoms and molecules. The study of hydrogen in quantum physics continues to be an active area of research and promises to yield new insights into the nature of matter and energy.

Helium is a chemical element with the symbol He and atomic number 2. It is the second lightest and second most abundant element in the universe, after hydrogen. Helium plays a crucial role in quantum physics due to its unique properties, which have led to some of the most important discoveries in this field. One of the most significant contributions of helium to quantum physics is its use in the study of superfluidity. Superfluidity is a quantum phenomenon in which a fluid flows with zero viscosity, meaning it can flow without any resistance. Helium-4, one of the isotopes of helium, exhibits superfluidity at very low temperatures, below 2.17 Kelvin. This has led to the study of the properties of superfluid helium-4, which has yielded important insights into quantum mechanics. The behavior of superfluid helium-4 is described by the Gross-Pitaevskii equation, which is a nonlinear partial differential equation that describes the macroscopic wave function of the superfluid. This equation has been used to study the behavior of superfluid helium-4, including its ability to flow without resistance and its ability to move through small openings in containers. Another important contribution of helium to quantum physics is its use in the study of quantum gases. Helium-4 and helium-3, the two isotopes of helium, are both bosons, which means they obey Bose-Einstein statistics. This makes them ideal for studying the behavior of bosons in quantum gases. One of the most important experiments in the study of quantum gases was the discovery of Bose-Einstein condensation. In this phenomenon, bosons are cooled to very low temperatures, causing them to occupy the same energy level and behave as a single quantum object. This was first observed in 1995 using rubidium atoms, and subsequent experiments have confirmed the phenomenon in other bosonic systems, including helium-4. Helium has also played a role in the development of quantum computing. Quantum computing is a field that aims to use the principles of quantum mechanics to develop computers that are more powerful than classical computers. One of the challenges of quantum computing is maintaining coherence in quantum systems, which can be disrupted by thermal fluctuations and other factors. Helium has been used in the development of cryogenic systems that can maintain low temperatures necessary for quantum computing. Helium-3 has also been used in the study of qubits, which are the basic units of information in quantum computing. In conclusion, helium plays a crucial role in quantum physics due to its unique properties. Its ability to exhibit superfluidity and its use in the study of quantum gases have led to some of the most important discoveries in this field. Helium has also played a role in the development of cryogenic systems for quantum computing. The study of helium in quantum physics continues to be an active area of research and promises to yield new insights into the nature of matter and energy.

Nitrogen is a chemical element with the symbol N and atomic number 7. It is a non-metal and is the most abundant element in Earth's atmosphere, making up around 78% of the air we breathe. Nitrogen plays an important role in quantum physics due to its unique properties, which have led to a number of important discoveries in this field. One of the most important contributions of nitrogen to quantum physics is its use in the study of solid-state physics. Nitrogen is commonly used as a cryogenic coolant, which can cool materials down to very low temperatures, often near absolute zero. This allows researchers to study the behavior of matter at the quantum level. One important application of cryogenics in solid-state physics is the study of superconductivity. Superconductivity is a phenomenon in which certain materials can conduct electricity with zero resistance at very low temperatures. Nitrogen is often used to cool these materials down to the temperatures at which superconductivity occurs.

Nitrogen is also important in the study of the quantum behavior of molecules. The nitrogen molecule, N2, is a diatomic molecule that is used in a variety of industrial processes. The electronic structure of the nitrogen molecule has been studied extensively using quantum mechanics. In particular, the nitrogen molecule is an example of a homonuclear diatomic molecule, meaning it is made up of two atoms of the same element. The study of homonuclear diatomic molecules has led to a better understanding of the relationship between the electronic structure of molecules and their physical properties. Nitrogen is also important in the study of quantum information. Quantum information is a field that aims to use the principles of quantum mechanics to process and transmit information more efficiently than classical computers. Nitrogen is used in the development of nitrogen-vacancy centers, which are defects in diamonds that can be used as qubits in quantum computers. In addition, nitrogen is used in the study of nitrogen-vacancy centers in biological systems. These centers can be used to study the properties of biological molecules, such as proteins and enzymes, at the quantum level. This has important applications in the development of new drugs and therapies. In conclusion, nitrogen plays an important role in quantum physics due to its unique properties. Its use in cryogenics has allowed researchers to study the behavior of matter at the quantum level, and its electronic structure has been studied extensively using quantum mechanics. Nitrogen is also important in the study of quantum information and has important applications in the study of biological systems. The study of nitrogen in quantum physics continues to be an active area of research and promises to yield new insights into the nature of matter and energy.

Lithium is a chemical element with the symbol Li and atomic number 3. It is a soft, silvery-white metal that is highly reactive and is the lightest of all metals. Lithium plays an important role in quantum physics due to its unique properties, which have led to a number of important discoveries in this field. One of the most important contributions of lithium to quantum physics is its use in the study of atomic physics. Lithium has only three electrons, making it one of the simplest atoms. This simplicity has allowed researchers to study the electronic structure of lithium using quantum mechanics. In particular, lithium is used in the study of

Bose-Einstein condensates. Bose-Einstein condensates are a state of matter that occurs when bosons, particles that obey Bose-Einstein statistics, are cooled to very low temperatures, causing them to occupy the same quantum state. This results in a macroscopic wave function that describes the entire system. Lithium is an ideal element for the study of Bose-Einstein condensates because it has two stable isotopes, lithium-6 and lithium-7, that can be used to create condensates with different properties. The study of these condensates has led to a better understanding of the behavior of quantum systems. Lithium is also important in the study of quantum magnetism. Quantum magnetism is a field that studies the magnetic properties of materials at the quantum level. Lithium has a strong spin-orbit interaction, meaning that the motion of its electrons is strongly influenced by their magnetic moment. This makes lithium an ideal element for the study of magnetic properties in quantum systems. Lithium has been used in the study of spinor Bose-Einstein condensates, which are condensates that exhibit both magnetic and superfluid properties. The study of these systems has led to a better understanding of the relationship between magnetism and superfluidity. Lithium is also important in the study of quantum computing. Lithium ions can be used as qubits, the basic units of information in quantum computing. Lithium ions have a long coherence time, meaning that they can maintain their quantum state for a relatively long period of time. In addition, lithium has been used in the study of quantum information in biological systems. Lithium is an important element in the study of lithium-ion channels, which are channels in cell membranes that allow ions to pass through. These channels play a crucial role in the transmission of nerve impulses and other biological processes.

In conclusion, lithium plays an important role in quantum physics due to its unique properties. Its simplicity makes it an ideal element for the study of atomic physics, and it is an important element in the study of Bose-Einstein condensates, quantum magnetism, and quantum computing. Lithium is also important in the study of quantum information in biological systems. The study of lithium in quantum physics continues to be an active area of research and promises to yield new insights into the nature of matter and energy.

Carbon is a chemical element with the symbol C and atomic number 6. It is a non-metallic element and is known for its ability to form strong covalent bonds with other elements. Carbon plays a crucial role in quantum physics due to its unique properties, which have led to numerous discoveries in this field. One of the most important contributions of carbon to quantum physics is its ability to form diamond, which is a crystalline form of carbon. Diamond has unique properties that make it an ideal material for quantum applications. For example, diamond is transparent, making it an ideal material for the construction of optical components such as lenses and prisms. Diamond is also an excellent thermal conductor, making it ideal for the construction of electronic components.

Diamond is also an important material for the study of quantum entanglement, which is a phenomenon in which two particles become entangled and share a quantum state. Diamond can be used to create nanoscale sensors that can detect the presence of single electrons. These sensors can be used to study the behavior of entangled particles and have led to a better understanding of the nature of quantum entanglement. Carbon is also important in the study of quantum computing. Carbon nanotubes, which are cylindrical structures made of carbon atoms, are an important material for the construction of quantum computers. Carbon nanotubes have unique electronic properties that make them an ideal material for the construction of qubits, the basic units of information in quantum computing. In addition, carbon is an important element in the study of graphene, which is a two-dimensional form of carbon. Graphene has unique electronic properties that make it an ideal material for the study of quantum transport, which is the study of the movement of electrons in a material. Graphene has been used to study the quantum Hall effect, which is a phenomenon in which the resistance of a material to the flow of electrons is quantized in integer multiples of a fundamental constant. Carbon is also important in the study of quantum chemistry, which is the study of chemical reactions at the quantum level. Carbon is an important element in the study of organic chemistry, which is the study of carbon-based molecules. Organic chemistry is crucial to the study of life and the development of new drugs and materials. In conclusion, carbon plays a crucial role in quantum physics due to its unique properties. Its ability to form diamond and carbon nanotubes has led to important discoveries in the study of quantum entanglement and quantum computing. Graphene, a two-dimensional form of carbon, has unique electronic properties that make it an ideal material for the study of quantum transport. Carbon is also important in the study of quantum chemistry and organic chemistry. The study of carbon in quantum physics continues to be an active area of research and promises to yield new insights into the nature of matter and energy.

Argon is a chemical element with the symbol Ar and atomic number 18. It is a noble gas that is known for its inertness and its ability to remain stable under most conditions. Despite its apparent lack of reactivity, argon has made important contributions to the field of quantum physics. One of the most important contributions of argon to quantum physics is its use as a standard for the measurement of time. Atomic clocks are devices that use the vibrations of atoms to measure time. The vibrations of atoms are very precise, and atomic clocks are able to measure time to within a few billionths of a second. Argon is used as a standard for the measurement of time because it has a well-defined atomic structure and is readily available. Argon is also important in the study of quantum mechanics, which is the study of the behavior of matter and energy at the atomic and subatomic level. Quantum mechanics is a fundamental theory that has led to many important discoveries in physics, such as the discovery of the electron and the development of the transistor. Argon has been used to study the properties of atoms and molecules at the quantum level. For example, argon has been used to study the properties of helium, which is another noble gas. Helium has a well-defined atomic structure and is used as a standard for the measurement of time in atomic clocks. The study of helium has led to a better understanding of the properties of other atoms and molecules at the quantum level. Argon has also been used in the study of superfluidity, which is a phenomenon in which certain liquids flow with zero viscosity at very low temperatures. Superfluidity is a quantum effect that is not fully understood, but it is believed to be related to the behavior of particles at the quantum level. Argon has been used as a model system to study the properties of superfluid helium, which is one of the most studied superfluids. In addition, argon has been used in the study of plasma physics, which is the study of the behavior of ionized gases. Plasma is a state of matter that is found in stars, lightning, and other high-energy phenomena. Argon is one of the most common gases used in plasma research because it is readily available and has a well-defined atomic structure.

In conclusion, argon has made important contributions to the field of quantum physics. Its use as a standard for the measurement of time has led to more precise measurements and has improved our understanding of the behavior of matter and energy at the atomic and subatomic level. Argon has also been used to study the properties of atoms and molecules at the quantum level, the phenomenon of superfluidity, and the behavior of ionized gases.

131

The study of argon in quantum physics continues to be an active area of research and promises to yield new insights into the nature of matter and energy.

Now you know the most fundamental elements that contributed to the field of quantum physics and what role they played as well as a general background of the element.

Chapter 10: New Discoveries

In this chapter we will talk about the new and upcoming discoveries and innovations that will advance our undersntading of quantum physics in the near future.

The kilogram is one of the seven base units in the International System of Units (SI) and is the unit of mass. Until recently, the kilogram was defined by the mass of the International Prototype of the Kilogram, a cylinder of platinum-iridium alloy, which was kept at the International Bureau of Weights and Measures in France. However, this definition was not stable over time, as the mass of the prototype could change due to various factors such as pollution, wear, and loss of material.

In 2019, the definition of the kilogram was revised, and it is now based on the Planck constant, a fundamental constant of nature in quantum mechanics. This new definition has been adopted to provide a more stable and universal definition of the kilogram that is independent of the International Prototype of the Kilogram.

The Planck constant, denoted by h, is a physical constant that relates the energy of a photon to its frequency. It is named after the German physicist Max Planck, who first introduced it in 1900 to explain the spectrum of radiation emitted by a blackbody.

The value of the Planck constant is approximately 6.62607015 × 10^-34 joule-seconds.

To understand how the Planck constant can be used to measure the kilogram, we need to look at the concept of mass-energy equivalence, which is expressed by Einstein's famous equation $E = mc^2$. This equation shows that mass and energy are equivalent and can be converted into each other. In other words, the mass of an object is related to its energy content.

The Planck constant is related to the energy of a photon by the equation $E = hf$, where f is the frequency of the photon. Using this equation, we can write the energy of a photon in terms of its wavelength, λ, as $E = hc/\lambda$, where c is the speed of light. We can also write the momentum of a photon as $p = h/\lambda$.

In 1973, the International System of Units (SI) introduced a new definition of the meter, which is based on the speed of light in vacuum. The speed of light is defined as exactly 299,792,458 meters per second. Using this definition, we can express the Planck constant in terms of the kilogram, meter, and second. The Planck constant can be written as h = 6.62607015 × 10^-34 joule-seconds = 6.62607015 × 10^-34 kg·m^2/s.

By defining the Planck constant as a fixed value, we can use it to define the kilogram in a more stable and universal way. The current definition of the kilogram is based on the Planck constant and the Avogadro constant, which relates the number of atoms or molecules in a substance to its mass. The kilogram is defined as the mass of a certain number of atoms of a particular isotope of silicon, where the number of atoms is fixed by the Avogadro constant. The

Planck constant is used to relate the mass of these atoms to a fixed value in terms of kilograms.

In summary, the measurement of the kilogram is now based on the Planck constant, a fundamental constant of nature in quantum mechanics. This new definition provides a more stable and universal definition of the kilogram that is independent of the International Prototype of the Kilogram. The Planck constant is related to the energy and momentum of photons, and by defining it as a fixed value, we can use it to relate the mass of atoms to a fixed value in terms of kilograms.

Heat transfer is the process of transferring thermal energy from one object to another. One method of heat transfer is through radiation, which occurs when electromagnetic waves are emitted from a hot object and absorbed by a cooler object. In a vacuum, heat transfer by radiation is the only method available. In this paper, we will explore the role of quantum mechanics in heat transfer through a vacuum.

In classical physics, it was believed that heat transfer occurred through the transfer of energy from one molecule to another. However, this model fails to explain heat transfer in a vacuum, where there are no molecules present. In 1900, Max Planck proposed a new theory of radiation, which stated that electromagnetic radiation is quantized, meaning that it is composed of discrete packets of energy, now known as photons.

The theory of radiation led to the development of quantum mechanics, which describes the behavior of particles on a microscopic scale. According to quantum mechanics, particles such as photons have both wave-like and particle-like properties, known as wave-particle duality. This duality means that particles can act as waves and interfere with each other, or they can act as particles and be detected as discrete entities.

In heat transfer through a vacuum, radiation occurs when a hot object emits photons, which travel through the vacuum and are absorbed by a cooler object. The transfer of energy occurs through the absorption of photons by the cooler object. The rate of heat transfer through radiation is proportional to the temperature difference between the hot and cool objects, and is given by the Stefan-Boltzmann law.

The Stefan-Boltzmann law states that the rate of heat transfer by radiation is proportional to the fourth power of the temperature difference between the two objects:

$$ \frac{dQ}{dt} = \sigma A (T_1^4 - T_2^4) $$

where $\frac{dQ}{dt}$ is the rate of heat transfer, σ is the Stefan-Boltzmann constant, A is the surface area of the objects, and T_1 and T_2 are the temperatures of the hot and cool objects, respectively.

The quantum nature of photons also has implications for the way heat transfer occurs. In a vacuum, there are no other particles present to interact with the photons, so they travel through the vacuum without being scattered or absorbed. However, the act of detection by a cooler object causes the photons to be absorbed, and their energy is transferred to the object in the form of heat. This process is stochastic, meaning that the absorption of photons occurs randomly over time.

In addition to the stochastic nature of heat transfer through a vacuum, there are also limitations to the precision with which we can measure the temperature of an object. This is due to the uncertainty principle of quantum mechanics, which states that the more precisely we know the position of a particle, the less precisely we can know its momentum, and vice versa.

This uncertainty also applies to the measurement of temperature, since the measurement of temperature involves the measurement of the position and momentum of the atoms and molecules that make up the object. The measurement of temperature is therefore subject to a fundamental uncertainty, known as the quantum limit of temperature measurement.

In order to overcome this limitation, scientists have developed a method of temperature measurement using the principles of quantum mechanics. This method involves the use of a device known as a quantum thermometer, which uses the interaction between a quantum system and a thermal system to measure temperature with high precision.

In conclusion, heat transfer through a vacuum is a fundamental process in thermodynamics and is described by the principles of quantum mechanics. The stochastic nature of heat transfer and the limitations of temperature measurement due to the uncertainty principle has important implications for the design of thermal systems and the measurement of temperature. The development of new methods of temperature

Quantum gravity is an area of theoretical physics that aims to reconcile two of the most successful theories of physics, quantum mechanics and general relativity. While quantum mechanics describes the behavior of particles on small scales, general relativity describes the behavior of gravity on large scales, such as the motions of planets and stars. However, these theories are incompatible with each other, and quantum gravity seeks to develop a theory that unifies them.

One of the major challenges in developing a theory of quantum gravity is the fact that the equations of general relativity involve continuous, smooth space-time, while the equations of quantum mechanics involve discrete, non-continuous values. This creates a problem when trying to apply quantum mechanics to the phenomenon of gravity, which is a continuous force.

One of the most promising approaches to quantum gravity is string theory. String theory proposes that the fundamental particles of the universe are not point-like objects, but rather tiny, one-dimensional objects called strings. These strings vibrate at different frequencies, giving rise to the different particles and forces we observe in the universe. String theory also proposes that space-time is not a continuous fabric, but rather a discrete, quantized structure.

Another approach to quantum gravity is loop quantum gravity. In loop quantum gravity, space-time is quantized into tiny, discrete loops. These loops are then used to construct a quantum theory of gravity. In this theory, gravity is described as the curvature of space-time caused by the presence of matter and energy.

Quantum gravity has several important implications for our understanding of the universe. One of the most significant is the idea that space-time itself may be quantized, meaning that it is made up of indivisible units or "atoms." This would fundamentally change our understanding of the nature of space and time.

Another important implication of quantum gravity is the existence of "quantum foam," a turbulent and constantly fluctuating sea of space-time at the smallest scales. This quantum foam is thought to be responsible for the phenomenon of "virtual particles," particles that appear and disappear in the vacuum of space.

Quantum gravity also has important implications for our understanding of the origins of the universe. One of the key challenges in cosmology is explaining the "Big Bang," the event that is thought to have created the universe as we know it. Quantum gravity may provide a way to unify the physics of the very small and the very large, and may shed new light on the nature of the universe at its inception.

In conclusion, quantum gravity is an area of physics that seeks to unify the two most successful theories of modern physics, quantum mechanics and general relativity. While there is currently no agreed-upon theory of quantum gravity, approaches such as string theory and loop quantum gravity show promise in providing a way to reconcile the discrete and continuous aspects of the universe. The implications of quantum gravity are far-reaching, potentially changing our understanding of the nature of space and time and shedding new light on the origins of the universe

Quantum tunneling is a fascinating phenomenon in quantum mechanics where a particle is able to pass through a potential barrier, despite not having enough energy to overcome it classically. This effect has been observed in a variety of physical systems, from solid-state electronics to nuclear fusion, and is an essential component of many quantum technologies. In this paper, we will explore the mathematics behind quantum tunneling, its physical implications, and some of its applications.

The wave function of a particle is governed by the Schrödinger equation, which describes the time evolution of the wave function. When a particle encounters a potential barrier, its wave function splits into two components, one of which reflects off the barrier and the other of which tunnels through the barrier. The probability of a particle tunneling through a potential barrier is given by the transmission coefficient, which is derived from the wave function using the boundary conditions at the potential barrier.

The transmission coefficient for a rectangular barrier of height V0 and width L is given by:

$$T = 4k1k2/(k1 + k2)^2$$

Where k1 = sqrt(2m(E-V0))/h and k2 = sqrt(2mE)/h are the wave numbers of the incident and transmitted waves respectively, m is the mass of the particle, E is its energy, and h is Planck's constant.

The tunneling probability decreases exponentially with increasing barrier width and height, making it an unlikely event for large barriers. However, at the nanoscale, tunneling can become a dominant mode of transport, allowing for the design of novel electronic devices, sensors, and quantum computers.

One of the most intriguing implications of quantum tunneling is its role in nuclear fusion reactions. In the Sun, hydrogen nuclei (protons) collide at extremely high temperatures and pressures, leading to fusion reactions that release large amounts of energy. However, the protons are repelled by a strong Coulomb barrier, which would prevent fusion from occurring classically. Quantum tunneling allows for the protons to tunnel through the barrier and come into close enough proximity for the strong nuclear force to bind them together.

Another application of quantum tunneling is in scanning tunneling microscopy (STM), a technique used to image surfaces at the atomic scale. In STM, a fine-tipped probe is brought very close to the surface being imaged, and the tunneling current between the probe and the surface is measured. By scanning the probe across the surface, an image of the surface can be reconstructed based on the variations in tunneling current. STM has revolutionized our understanding of surface structure and has led to the development of new materials and technologies.

In addition to nuclear fusion and STM, quantum tunneling has a wide range of applications in modern technology. One example is in flash memory, where electrons are able to tunnel through a thin insulating layer (known as a tunnel oxide) to store and retrieve data. Another application is in tunnel diodes, which are used as high-frequency electronic oscillators and amplifiers. Tunnel diodes rely on the resonant tunneling effect, where electrons are able to tunnel through a narrow potential barrier at specific voltages, leading to a sharp increase in current.

Quantum tunneling is a fascinating phenomenon that arises from the wave-like nature of particles in quantum mechanics. The mathematics behind tunneling is based on the Schrödinger equation and the transmission coefficient, which describe the probability of a particle tunneling through a potential barrier. Quantum tunneling has important physical implications, such as its role in nuclear fusion reactions and scanning tunneling microscopy, and has a wide range of technological applications in fields such as electronics and materials science.

Conclusion

Quantum physics, also known as quantum mechanics, is a fundamental theory in physics that describes the behavior of matter and energy at the atomic and subatomic level. It is based on the principle of quantum superposition, which states that particles can exist in multiple states or positions at the same time until they are observed or measured. This concept is fundamentally different from classical physics, which assumes that particles have definite positions and properties. Quantum mechanics was developed in the early 20th century by physicists such as Max Planck, Albert Einstein, and Erwin Schrödinger. It has led to many important discoveries and applications, including the development of the transistor, the laser, and the theory of superconductivity. One of the key principles of quantum mechanics is the uncertainty principle, which states that certain pairs of properties of a particle, such as its position and momentum, cannot be measured simultaneously with complete accuracy. This means that the more precisely one property is measured, the less precisely the other property can be known. Another important principle of quantum mechanics is wave-particle duality, which states that particles can exhibit both wave-like and particle-like behavior depending on how they are observed or measured. For example, light can behave like a wave or a particle depending on the experiment. Quantum mechanics also includes the concept of entanglement, in which particles can become linked in such a way that the state of one particle is dependent on the state of another particle, regardless of the distance between them. This concept has important implications for communication and cryptography. In addition, quantum mechanics includes the idea of quantization, which states that certain properties of particles, such as their energy levels, can only take on discrete values. This concept is important in the study of atomic and molecular spectra, which can be used to identify the elements and molecules present in a sample. Quantum mechanics is a complex and abstract theory that is not fully understood, and it continues to be an active area of research. Its principles have led to many important discoveries and applications, and they have challenged our understanding of the nature of reality at the fundamental level.

Nicholas Tassy-Becz

Thank you for reading and be sure to check out new and upcoming books from Nicholas Tassy-Becz like

How the World Spins Round

Works Cited

"●." *YouTube*, 10 August 2022, http://ocw.mit.edu/8%E2%80%9304S13. Accessed 21 February 2023.

"●." *YouTube*, 10 August 2022, http://nptel.ac.in/courses.php. Accessed 21 February 2023.

"●." *YouTube*, 10 August 2022, http://www.physics.org/explorelink.asp?id=5446. Accessed 21 February 2023.

"●." *YouTube*, 10 August 2022, https://ocw.mit.edu/search/?d=Physics&s=department_course_numbers.sort_coursenum. Accessed 21 February 2023.

"●." *YouTube*, 10 August 2022, http://www.physics.org/explorelink.asp?id=5231. Accessed 21 February 2023.

Cassidy, David C. *Beyond Uncertainty: Heisenberg, Quantum Physics, and the Bomb.* Bellevue Literary Press, 2010.

Deffner, Sebastian, and Steve Campbell. *Quantum Thermodynamics: An Introduction to the Thermodynamics of Quantum Information.* Morgan & Claypool Publishers, 2019.

Dick, Rainer. *Advanced Quantum Mechanics: Materials and Photons*. Springer

International Publishing, 2021.

Gillet, Jean-Michel. *Application-driven Quantum and Statistical Physics: A Short

Course for Future Scientists and Engineers. Transitions. Volume 3*. World

Scientific Publishing UK Limited, 2020.

Griffiths, David J. *Introduction to Quantum Mechanics*. Cambridge University Press,

2017.

Griffiths, David J., and Darrell F. Schroeter. *Introduction to Quantum Mechanics*.

Cambridge University Press, 2018.

Holzner, Steven. *Quantum Physics Workbook For Dummies*. Wiley, 2010.

Raymer, Michael G. *Quantum Physics: What Everyone Needs to Know*. Oxford

University Press, 2017.

Stehle, Philip. *Order, Chaos, Order: The Transition from Classical to Quantum

Physics*. Oxford University Press, 1994.

The Quantum Frontier

www.ingramcontent.com/pod-product-compliance
Lightning Source LLC
Chambersburg PA
CBHW021951170526
45157CB00003B/934